Renewable Energy: Research, Development and Policies

Renewable Energy Development on Federal Lands

RENEWABLE ENERGY: RESEARCH, DEVELOPMENT AND POLICIES

Additional books in this series can be found on Nova's website under the Series tab.

Additional E-books in this series can be found on Nova's website under the E-book tab.

RENEWABLE ENERGY: RESEARCH, DEVELOPMENT AND POLICIES

RENEWABLE ENERGY DEVELOPMENT ON FEDERAL LANDS

CARL POLAT
EDITOR

nova publishers
New York

For permission to use material from this book please contact us:
Telephone 631-231-7269; Fax 631-231-8175
Web Site: http://www.novapublishers.com

Additional color graphics may be available in the e-book version of this book.

Library of Congress Cataloging-in-Publication Data

ISBN: 978-1-62618-764-1

Published by Nova Science Publishers, Inc. †New York

CONTENTS

PREFACE

Concerns over reliance on imported oil and greenhouse gas emissions from fossil fuel use have led to increased interest in producing electricity from renewable sources, including wind, solar, and geothermal energy. Because federal lands, including those managed by the Departments of Agriculture and Interior, encompass areas with high renewable energy potential, interest has increased in permitting such activity on those lands. EPAct 2005 includes several provisions intended to increase renewable energy development on federal lands, including goals for approving renewable energy projects. This book examines the status of renewable energy permitting on federal land, including time frames for processing permits applied for since 2005; and the actions federal land management agencies have taken to facilitate this development.

Chapter 1 – Concerns over reliance on imported oil and greenhouse gas emissions from fossil fuel use have led to increased interest in producing electricity from renewable sources, including wind, solar, and geothermal energy. Because federal lands, including those managed by the Departments of Agriculture and the Interior, encompass areas with high renewable energy potential, interest has increased in permitting such activity on those lands. EPAct 2005 includes several provisions intended to increase renewable energy development on federal lands, including goals for approving renewable energy projects. GAO was asked to look at (1) the status of renewable energy permitting on federal land, including time frames for processing permits applied for since EPAct 2005; (2) actions federal land management agencies have taken to facilitate renewable energy development on federal land, particularly since the passage of EPAct 2005; and (3) factors affecting renewable energy development on federal land. To conduct this work, GAO reviewed laws, regulations, and policies; interviewed agency and industry officials; and surveyed BLM staff responsible for processing applications for renewable energy permits on federal lands.

GAO is not making any recommendations in this report. In commenting on a draft of this report, the Department of Agriculture concurred with its findings, while the Departments of Energy and the Interior had no comments.

Chapter 2 – The authors evaluated the Bureau of Land Management's (BLM) renewable energy activities and found that BLM is poised for a massive expansion in renewable energy construction and development of wind and solar projects. At the time of the authors' review, BLM had 29 of 31 authorized wind projects in operation. It also had 137 authorized wind-testing projects that could transition into full development projects, and an additional 47 pending applications. BLM had five of nine authorized solar projects just beginning construction, and another 103 applications pending review.

Chapter 3 – Recent concerns over energy supply and pricing have led some to look increasingly to federal lands as a potential energy source. This report explains the legal framework for energy leasing and permitting for onshore lands subject to the control of the federal government. The report first reviews the laws and regulations affecting leasing of federal lands for exploration and production of oil, natural gas, and coal, as well as the permits that lessees must obtain in order to explore for and produce these resources. This leasing process has evolved over the last century under the framework established by the Mineral Leasing Act of 1920 (MLA). Oil, natural gas, and coal leasing and production on federal land pursuant to this act are currently overseen by the Bureau of Land Management (BLM), an agency within the U.S. Department of the Interior (DOI). Federal lands identified in the area's Resource Management Plan (RMP) as amenable to oil, coal, or natural gas exploration and production may be leased by BLM, so long as such activities in that area are not prohibited by statute or regulation. Such lands are usually leased to the highest bidder as determined by competitive auction. Leaseholders are generally required to pay both rental fees and royalties (a percentage of the value of produced oil, natural gas, or coal) to the U.S. government. The report also addresses existing laws and regulations that govern the use of federal lands for renewable energy projects, including geothermal, wind, and solar energy. BLM oversees permitting for these projects. Geothermal projects are leased in accordance with the requirements of the Geothermal Steam Act of 1970. That act functions similarly to the MLA; lands that are amenable to geothermal projects are leased to the highest qualified bidder. In contrast, wind and solar projects on federal lands are not authorized by leases, but rather by obtaining a right-of-way from BLM. These rights-of-way are issued pursuant to the requirements of Title V of the Federal Land Policy and Management Act (FLPMA), and holders of these rights-of-way must make monthly rental payments to the U.S. government. This framework provides BLM and the federal government with flexibility to use federal lands to help satisfy the nation's energy needs, while generating revenue for the federal government and protecting environmentally sensitive areas.

Chapter 4 – The BLM manages 20.6 million acres of public lands with wind potential. The BLM's Lands and Realty Management program has authorized 32 wind energy development projects with a current total installed capacity of 581 megawatts, enough to supply the power needs of over 175,000 homes. In addition, the BLM has approved 170 wind energy site testing authorizations.

A Programmatic Environmental Impact Statement (EIS) relating to the development of wind energy on the public lands was completed in June 2005. This EIS provides an analysis of the development of wind energy projects on the public lands in the West. In conjunction with the publication of this EIS, the BLM amended 52 land use plans to allow for the use of applicable lands for wind energy development. BLM offices are able to use this EIS as an aid in analyzing impacts for specific applications for the use of public lands for wind energy use. The BLM issued a wind energy policy in December 2008 to provide guidance on best management practices (BMPs); measures to mitigate potential impacts on birds, wildlife habitat, and other resource values; and guidance on administering wind energy authorizations.

Chapter 5 – The BLM manages 20.6 million acres of public lands with wind potential. The BLM's Lands and Realty Management program has authorized 32 wind energy development projects with a current total installed capacity of 581 megawatts, enough to supply the power needs of over 175,000 homes. In addition, the BLM has approved 170 wind energy site testing authorizations.

A Programmatic Environmental Impact Statement (EIS) relating to the development of wind energy on the public lands was completed in June 2005. This EIS provides an analysis of the development of wind energy projects on the public lands in the West. In conjunction with the publication of this EIS, the BLM amended 52 land use plans to allow for the use of applicable lands for wind energy development. BLM offices are able to use this EIS as an aid in analyzing impacts for specific applications for the use of public lands for wind energy use. The BLM issued a wind energy policy in December 2008 to provide guidance on best management practices (BMPs); measures to mitigate potential impacts on birds, wildlife habitat, and other resource values; and guidance on administering wind energy authorizations.

Chapter 6 – The BLM has the delegated authority for leasing 245 million acres of public lands (including 104 million acres of National Forest managed by the U.S. Forest Service) with geothermal potential in 11 western States and Alaska. The BLM presently manages 818 geothermal leases, with 59 geothermal leases in producing status generating about 1,275 megawatts of installed geothermal energy on public lands. This amounts to over 40 percent of U.S. geothermal energy capacity and supplies the electrical needs of about 1.2 million homes.

In May 2007, the Department of the Interior published final regulations on geothermal energy production on public lands requiring more competitive leasing, offering simplified royalty calculations, and sharing $4 million per year in current royalties with counties where production occurs.

In: Renewable Energy Development ... ISBN: 978-1-62618-764-1
Editor: Carl Polat

Chapter 1

RENEWABLE ENERGY: AGENCIES HAVE TAKEN STEPS AIMED AT IMPROVING THE PERMITTING PROCESS FOR DEVELOPMENT ON FEDERAL LANDS*

United States Government Accountability Office

WHY GAO DID THIS STUDY

Concerns over reliance on imported oil and greenhouse gas emissions from fossil fuel use have led to increased interest in producing electricity from renewable sources, including wind, solar, and geothermal energy. Because federal lands, including those managed by the Departments of Agriculture and the Interior, encompass areas with high renewable energy potential, interest has increased in permitting such activity on those lands. EPAct 2005 includes several provisions intended to increase renewable energy development on federal lands, including goals for approving renewable energy projects. GAO was asked to look at (1) the status of renewable energy permitting on federal land, including time frames for processing permits applied for since EPAct 2005; (2) actions federal land management agencies have taken to facilitate renewable energy development on federal land, particularly since the passage

* This is an edited, reformatted and augmented version of the Highlights of GAO-13-189, a report to the Ranking Member, Committee on Natural Resources, House of Representatives, dated January 2013.

of EPAct 2005; and (3) factors affecting renewable energy development on federal land. To conduct this work, GAO reviewed laws, regulations, and policies; interviewed agency and industry officials; and surveyed BLM staff responsible for processing applications for renewable energy permits on federal lands.

GAO is not making any recommendations in this report. In commenting on a draft of this report, the Department of Agriculture concurred with its findings, while the Departments of Energy and the Interior had no comments.

What GAO Found

Since passage of the Energy Policy Act of 2005 (EPAct 2005), federal land management agencies—primarily the Department of the Interior's Bureau of Land Management (BLM)—have received hundreds of applications for utility- scale renewable energy projects and authorized 25 projects: 7 wind, 10 solar, and 8 geothermal projects. Applications for the majority of projects were withdrawn by the applicants or denied by BLM because of insufficient information. Applications for about one-fourth of the projects are still pending with the agencies. Time frames for permitting wind and solar projects ranged from 1.5 to 4 years from receipt of the initial application to approval of the project, with time frames decreasing for applications submitted in later years. For geothermal projects, permitting time frames ranged from 1 to 4 years from receipt of the initial application to approval for construction. In all, for projects applied for since EPAct 2005, BLM has authorized projects with the capacity to generate a total of about 5,450 megawatts of electricity, contributing to the act's goal of approving projects capable of generating 10,000 megawatts of electricity on public lands by 2015.

Federal land management agencies have taken several steps to foster renewable energy development on federal lands since EPAct 2005. Specifically, these agencies have developed or revised policies aimed at, among other things, improving the renewable energy permitting process, formalized coordination within and across agencies and with state and local governments, and devoted increased resources to processing applications for renewable energy permits. One of BLM's most comprehensive actions was the completion of programmatic environmental impact statements for renewable energy development, intended to streamline the permitting process. The agencies also took steps to improve coordination through regularly established meetings and development of memorandums of understanding between federal and state agencies. They also added staff and increased funding for this development. For example, BLM tripled its staff devoted to processing wind and solar energy applications. To help ensure that its actions are achieving their intended purposes, BLM issued an instruction memorandum in December 2012 aimed at increasing the efficiency and effectiveness of its renewable energy permitting process.

According to BLM respondents to a GAO questionnaire, industry representatives, and others GAO interviewed, many factors affect the pace of renewable energy development on federal lands. Some of these factors are specifically tied to the agencies' permitting processes, primarily BLM's. For example, respondents cited effective coordination among the involved parties and the amount of resources the agency can devote to permitting as factors that facilitated the permitting process. On the other hand, they often cited problems with the quality of applications received as a factor that may hinder or slow the permitting process. Respondents also cited a number of factors outside of permitting agencies' control that can affect the pace of renewable energy development, such as access to transmission lines (which are often scarce in areas where renewable energy is abundant) and competition from electricity generated using conventional energy sources, such as natural gas.

ABBREVIATIONS

BLM	Bureau of Land Management
EPAct 2005	Energy Policy Act of 2005
NEPA	National Environmental Policy Act

January 18, 2013

The Honorable Edward J. Markey
Ranking Member
Committee on Natural Resources
House of Representatives

Dear Mr. Markey:

American families and businesses depend on electricity to power their lives, businesses, and the economy. Concerns over reliance on imported oil and greenhouse gas emissions from burning fossil fuels to generate power have increased interest in producing electricity from renewable sources, including wind, solar, and geothermal energy. In the coming decades, the Energy Information Administration in the Department of Energy projects the strongest growth in the domestic energy sector will come from renewable energy. The federal government is uniquely positioned to affect the development of renewable energy sources, in part through its land management activities. For example, the Department of the Interior's Bureau of Land Management (BLM) manages approximately 248 million surface acres, or about one-eighth of all lands in the United States, and estimates that 20.6 million acres have commercial wind energy generation potential and more than 20 million acres have commercial solar energy generation potential.

About 95 percent of federal land is managed by four land management agencies—BLM, the Fish and Wildlife Service, and the National Park Service in the Department of the Interior and the Forest Service in the Department of Agriculture. The agencies manage this land for various purposes. BLM and the Forest Service manage land for multiple uses, such as recreation, timber, and fish and wildlife, and have much of the responsibility for permitting renewable energy development on federal lands. In contrast, the Fish and Wildlife Service and the National Park Service manage land primarily to conserve fish and wildlife and their habitats and park resources, respectively.

Because areas of federal land have a high potential for producing renewable energy, interest has increased in permitting such production on those lands over the past decade. For example, through the Energy Policy Act of 2005 (EPAct 2005), Congress directed the Secretary of the Interior to "seek to have approved non-hydropower renewable energy projects located on public lands with a generation capacity of at least 10,000 megawatts of electricity" by 2015.[1] In 2009 the Secretary issued an order establishing renewable energy as a priority for the department and also establishing a task force to, among other tasks, develop a strategy to increase the development and transmission of renewable energy from appropriate areas on public lands.[2]

Although renewable energy is generally cleaner to produce than energy from conventional sources, such as coal- or gas-fired power plants, renewable energy projects can have significant environmental impacts. Developing renewable energy projects on a scale large enough to generate substantial electric power (i.e., on a utility scale) can affect the environment across hundreds, even thousands, of acres of federal land and, depending on the energy source, may preclude uses of those acres for other purposes. In addition, these projects, like other development projects, can affect the surrounding environment: for example, water used in some solar technologies may diminish scarce groundwater in the arid Southwest, or the turning rotors of a wind turbine may kill birds and bats. Accordingly, part of the federal permitting process for renewable energy development on BLM- and Forest Service-managed lands involves not only those agencies but also the Fish and Wildlife Service and National Park Service in assessing potential environmental effects.

In light of interest in renewable energy development on federal land, you asked us to review federal agencies' processes and time frames for permitting renewable energy projects. Our objectives were to examine (1) the status of renewable energy permitting on federal land, including time frames for processing permits applied for since EPAct 2005; (2) actions the agencies have taken to facilitate renewable energy development on federal land, particularly since the passage of EPAct 2005; and (3) factors affecting renewable energy development on federal land.

To determine the status of renewable energy permitting, we administered a questionnaire to BLM officials nationwide, covering all permit applications for onshore wind, solar, and geothermal energy projects submitted to BLM—the agency responsible for permitting almost all applications submitted for renewable energy development on federal lands—from enactment of EPAct 2005 (August 8, 2005) through May 2012. We analyzed information obtained through the questionnaire regarding the status and time frames associated with each application and the factors affecting its processing. We received a 100 percent response rate to this questionnaire. In addition, we conducted semistructured interviews with officials in the nine Forest Service regional offices to determine the status and time frames of renewable energy permitting on Forest Service-managed lands. To determine actions taken to facilitate renewable energy development, we reviewed relevant laws, regulations, and agency policies and guidance. We interviewed headquarters officials from the Department of the Interior and its four land management agencies—BLM, the Bureau of Indian Affairs, the Fish and Wildlife Service, and the National Park Service—as well as from the Department of Agriculture's Forest Service. We also interviewed BLM officials from four state offices;[3] Fish and Wildlife Service and National Park Service officials from selected regions and field units;[4] officials from all nine Forest Service regions; and officials in the wind, solar, and geothermal research programs at the Department of Energy's National Renewable Energy Laboratory in Golden, Colorado. We obtained BLM funding and staffing data from the agency's database, and we assessed the reliability of these data by reviewing the agencies' internal controls of their data systems and interviewing agency officials; we found these data to be sufficiently reliable for the purposes of this report. To determine factors affecting renewable energy development, we used the results of our questionnaire and also interviewed representatives from industry and environmental groups to obtain their perspectives on the time frames and factors associated with the permitting process, and renewable energy development in general. Appendix I provides a more detailed description of our scope and methodology.

We conducted this performance audit from January 2012 to January 2013 in accordance with generally accepted government auditing standards. Those standards require that we plan and perform the audit to obtain sufficient, appropriate evidence to provide a reasonable basis for our findings and conclusions based on our audit objectives. We believe that the evidence obtained provides a reasonable basis for our findings and conclusions based on our audit objectives.

BACKGROUND

Renewable energy technologies generate electricity, fuels, or heat through the use of resources that are continually replenished, such as wind, sunlight, and naturally occurring underground steam and heat.[5] Development of utility-scale renewable energy projects on federal land occurs primarily on lands managed by BLM and, to a lesser extent, on those managed by the Forest Service. Nearly all of the approximately 248 million surface acres of federal land managed by BLM are located in 11 western states and Alaska. BLM is also responsible for managing resources (e.g., geothermal, oil, and gas resources) lying on or beneath federal lands and beneath private lands for which the federal government retains mineral rights—amounting to approximately 700 million subsurface acres altogether. The Forest Service is responsible for managing 193 million acres of forests and grasslands, primarily in the western states but also throughout the country. The Fish and Wildlife Service and National Park Service manage about 90 million and 80 million acres, respectively, but under legislation that generally precludes the development of large-scale renewable energy projects.[6] Another Interior agency, the Bureau of Indian Affairs, works with Indian tribes to develop renewable energy on tribal lands.[7]

Renewable energy projects on federal lands date back decades; for example, utility-scale geothermal projects have operated on both BLM- and

Forest Service-managed lands since the 1970s, and several utility- scale wind projects have been operating on BLM-managed lands, primarily in California, since the early 1980s.[8] In May 2012, the first utility- scale solar power plant began operating on federal lands, specifically on BLM-managed land. No utility-scale wind or solar development is operating at present on Forest Service-managed lands. In general, over half the nation's electricity generated from geothermal resources comes from resources on federal lands, and about 1 percent of the nation's electricity generated from wind and solar energy comes from resources on federal lands.

Types of Renewable Energy and Technologies Used for Development

Electricity generation using wind, solar, and geothermal energy can range from small-scale production—for example, rooftop solar panels on a home or geothermal resources heating a greenhouse—to utility-scale production of hundreds of megawatts of electricity. Figure 1 shows examples of utility-scale facilities representing each of these three energy types.

The most common technologies for utility-scale wind, solar, and geothermal development include the following:

- *Wind energy.* Wind farms comprise a number of turbines built close together to produce utility-scale wind power. Horizontal-axis turbines, the most common, constitute nearly all utility-scale turbines in the United States.[9] To generate electricity, horizontal-axis turbines capture the wind's energy with two or three propellerlike blades mounted on a rotor sitting atop a tower. A smaller utility-scale wind farm may have 6 turbines on 100 acres and generate about 3 megawatts, and a larger wind farm may have over 100 turbines on about 10,000 acres and generate over 300 megawatts. Siting wind farms in appropriate locations is important because they can fragment wildlife habitat and plant communities, and the turbines and blades themselves may kill birds and bats, make noise, and mar views.
- *Solar energy.* Multiple technologies exist for utility-scale solar power. Solar photovoltaic technologies convert energy from sunlight directly into electricity, using arrays of solar panels. Concentrating solar power technologies use mirrors to focus the sun's energy to heat water

or other fluids; these fluids in turn create steam that powers a conventional turbine generator to produce electricity. An average solar plant, whether photovoltaic or concentrating, requires about 3 to 8 acres for every megawatt of generating capacity. For example, a small utility-scale solar plant occupying about 400 acres may generate 45 megawatts, and a large facility occupying over 7,000 acres may generate 1,000 megawatts. Selecting appropriate areas for the development of solar projects is essential given that such development generally precludes other uses of the same land because the surface area is graded before installation, and components are installed relatively close to one another. In the southwestern United States, where the potential for solar energy development is greatest, solar installations can affect habitat needed by various species, including those listed as threatened or endangered under the Endangered Species Act, such as the desert tortoise. In addition, some solar technologies use large amounts of water, which can be problematic in desert environments.

- *Geothermal energy*. Geothermal power plants extract geothermal fluids—hot water, brines, and steam—from the earth by drilling wells to depths of up to 10,000 feet.[10] These fluids are then used to create a vapor that can power a turbine generator to produce electricity, with only the highest-temperature geothermal resources—generally above 200 degrees Fahrenheit—suitable for electricity generation. Geothermal power plants typically generate from 30 to 120 megawatts, and an individual power plant often uses resources brought to the surface through multiple wells. Mitigating for harmful effects is important in geothermal planning because geothermal operations can emit dangerous gases, and geothermal drilling operations may contaminate groundwater.

However it is generated, electricity from utility-scale renewable energy development is sold to utilities and—like electricity generated by more conventional sources such as coal- or gas-fired power plants—is conveyed to consumers through transmission lines. Some of the best locations for wind and solar development, however, are in remote areas far from consumers, and transmission lines to carry the power are not always readily available in these remote areas. In addition, electricity generated from wind and solar development differs from geothermal and conventional sources because it is available intermittently, rather than continuously.

Source: U.S. Department of Energy.

Figure 1. Examples of Utility-Scale Wind, Solar, and Geothermal Technologies.

Agency Permitting Processes for Renewable Energy Development

In managing their lands, BLM and the Forest Service are required by the Federal Land Policy and Management Act of 1976 and the National Forest Management Act, respectively, to develop or revise land use plans for the areas they manage, providing for multiple uses, such as recreation, timber, and fish and wildlife, and natural scenic, scientific, and historical values. All land management actions, including renewable energy development, must conform to the approved land use plan governing the land management unit—such as a national forest—where the action is to take place. Consequently, renewable energy development can occur only where it is consistent with the applicable land use plan, which in some cases may require changing the plan. Plan revisions undergo extensive environmental analysis before being finalized.

Wind and solar energy projects on federal land require permits from the relevant land management agency through the issuance of rights-of- way.[11] A right-of-way is an authorization to a qualified individual, business, or government entity to use a specific area of federal land for a specific amount of time for a certain purpose and with certain restrictions. Geothermal activities are administered by BLM for all federal lands (regardless of whether the surface land is managed by BLM or another agency) through the issuance first of leases and then of subsequent permits for drilling operations and other activities—similar to the process used for oil and natural gas.[12]

Permitting for Wind and Solar Energy Development

The Federal Land Policy and Management Act of 1976 authorizes BLM and the Forest Service to issue rights-of-way over federal lands for a variety of purposes, including systems for generating, transmitting, and distributing electric energy.[13] For new projects that may have a significant impact on the environment, the act requires applicants to submit a plan of construction, operation, and rehabilitation for the right-of-way that complies with applicable laws and regulations. An agency may issue a right-of-way only after the applicant has demonstrated that it has the technical and financial capability to construct the project for which the right-of-way is requested. Each right-of-way must include terms and conditions that, among others, protect the environment, federal property and economic interests, and the public interest. The holder of a right-of- way must generally pay its fair market value annually in advance and must provide a bond ensuring that the holder can perform the obligations—such as reclamation—required by the terms of the right-of- way.

BLM's permitting process for wind or solar energy projects comprises several steps prior to right-of-way issuance, as follows:

- A potential project developer contacts the BLM office responsible for land where a right-of-way is sought and obtains a standard application form.
- Preapplication meetings take place between the potential applicant and appropriate BLM field office staff to discuss the application form and its requirements, the general project proposal, land use planning in the area, and potential land use constraints.
- Once an application has been submitted, BLM reviews it to determine whether it is complete. A complete application must include a statement of technical and financial capability to construct, operate, maintain, and terminate the system for which the right-of-way is being requested, along with a project description—called a plan of development—sufficiently detailed for BLM to evaluate the proposed project's appropriateness and feasibility.[14]
- BLM checks to ensure that the proposed use conforms with BLM land use plans and that no apparent conflicts exist, such as other valid rights for the same lands requested in the application.
- BLM establishes a cost-recovery fee based on the amount of time BLM estimates it will take to process the application and issue a decision and collects this fee from the applicant.[15]
- BLM begins processing and evaluating the application, including associated environmental reviews under the National Environmental Policy Act (NEPA),[16] Endangered Species Act, and others.

An approved application generally results in an authorized right-of-way, although BLM generally issues a notice to proceed before the holder of the right-of-way can begin construction. On the other hand, BLM is authorized to deny applications for various reasons—for example, if the project is not consistent with the relevant land use plan or if applicants do not provide BLM with sufficient information.[17] At any point in the process, an applicant may choose to withdraw the application.

The Forest Service issues permits for wind and solar energy development under its "special use" permit process.[18] A potential project developer files a project proposal, which must contain information such as a project description and evidence of the applicant's technical and financial capability to carry out the project. Proposals are initially screened to ensure in part that the proposed project is consistent with applicable laws, regulations, and relevant land use plans. A second screening then examines whether the applicant has demonstrated sufficient technological and financial capability. If these conditions are met, the potential project developer may submit the proposal as a formal application, which is then analyzed under NEPA. If the project is approved, a special use permit is awarded to the applicant. Special use permits must include terms and conditions that protect the environment and require compliance with federal and state air and water quality laws, as well as with any state environmental and facility siting standards that are more stringent than federal standards. Like BLM, the Forest Service can deny applications for various reasons, including if the proposed project is inconsistent with the relevant land use plan or if the applicant fails to demonstrate sufficient technical or financial capability.

Permitting for Geothermal Energy Development

Geothermal activities are authorized through the issuance of leases and several subsequent permits and approvals to drill fluid minerals under the Geothermal Steam Act of 1970, as amended. To explore and develop geothermal resources on federal lands, developers generally must first obtain a federal lease from BLM.[19] Once a lease is obtained, a potential developer undergoes three primary phases seeking approval of plans and permits to explore and then develop geothermal resources and, ultimately, to construct a geothermal power plant. These plans and permits include the following: (1) an exploration plan, which describes the overall process for drilling and testing for geothermal resources; (2) one or more geothermal drilling permits, which allow an applicant to drill wells to confirm and produce geothermal resources at a given location; and (3) a utilization plan, which describes the proposed power plant and infrastructure needed to generate electricity from geothermal resources and provides much of the information needed to permit the power plant itself. A construction permit and site license must also be obtained before geothermal plant construction may begin. These three phases of geothermal permitting can be set in motion concurrently, but generally the environmental approval for the exploration phase is completed before applicants apply for geothermal drilling permits and subsequent construction permits. Applications for geothermal drilling permits can also be submitted after construction of a plant is completed if the developer determines that a geothermal field may benefit from additional wells.

Key Environmental Laws Affecting the Permitting Process

As with many other land management activities, the agencies must comply with key environmental laws—including, among others, NEPA, the Endangered Species Act, and the National Historic Preservation Act—when amending land use plans and approving permits for renewable energy development. Compliance with these laws can involve considerable time spent by agencies and project developers on environmental analysis and consultation with multiple agencies.

- *National Environmental Policy Act of 1969.*[20] Enacted in 1970, NEPA has as its purpose, among others, to promote efforts to prevent or eliminate damage to the environment. NEPA requires an agency to prepare a detailed statement on the environmental impacts of any "major federal action" significantly affecting the environment. Regulations promulgated by the Council on Environmental Quality[21] implementing NEPA generally require an agency to prepare either an environmental assessment[22] or an environmental impact statement.[23] Agencies may prepare an environmental assessment to determine whether there is a significant potential impact on the environment, which would necessitate the preparation of an environmental impact statement. If the agency determines in its environmental assessment that no significant environmental impacts will occur from the proposed action, then it prepares a finding of no significant impact. If the agency issues an environmental impact statement, it must also issue a record of decision describing the agency's decision; identifying all alternatives the agency considered; and stating whether all practicable means to avoid or minimize environmental harm from the alternative selected have been adopted and, if not, why not. Environmental impact statements can be developed at either a programmatic level—where larger-scale, combined and cumulative effects can be evaluated and where overall management objectives, such as road access and use, are defined—or a project level, where the effects are evaluated of a particular project in a specific place at a particular time. Programmatic environmental impact statements can reduce the environmental analysis needed for individual projects proposed in the area covered by a programmatic statement.[24] In addition, land use plan revisions typically involve the preparation of an environmental impact statement, and programmatic statements have also enabled agencies to amend multiple land use plans at the same time (e.g., land use plans for areas served by multiple BLM field offices), thereby reducing the NEPA review typically required when unit-specific land use plans are amended.

- *Endangered Species Act of 1973.*[25] The purpose of the Endangered Species Act is to conserve threatened and endangered species and the ecosystems upon which they depend. Under section 7 of the act, federal agencies must ensure that any action they authorize, fund, or carry out is not likely to jeopardize the continued existence of a species protected under the act. To fulfill this responsibility, the agencies must, under some circumstances, formally consult with the Fish and Wildlife Service when their actions may affect listed species or habitat identified as critical to the species' survival. The consultation usually ends with the issuance of a biological opinion by the Fish and Wildlife Service; this opinion may specify protective measures intended to minimize the project's impact on the species. For example, on the basis of a biological opinion, BLM may require a wind project to incorporate techniques reducing the turbines' potential to harm species protected under the act.
- *National Historic Preservation Act of 1966.*[26] The National Historic Preservation Act provides for the protection of historic properties. For all projects receiving federal funds or a federal permit, section 106 of the act requires federal agencies to take into account a project's effect on any historic property, including, for example, areas of traditional

religious and cultural importance to an Indian tribe.[27] In accordance with regulations implementing the act, agencies must consult with relevant federal, state, and tribal officials to determine whether a project or activity has the potential to affect historic properties.

Other key federal laws affecting the permitting of renewable energy projects in particular include the Migratory Bird Treaty Act[28] and the Bald and Golden Eagle Protection Act.[29] Each act generally makes it unlawful to "take" (defined to include, among other actions, hunting, trapping, or killing) a bird, its nest, or eggs covered by the act unless a permit has been issued under specific circumstances.

In addition to federal land management agency requirements, project applications are also subject to relevant state and local requirements, which vary from state to state. For example, in California, utility-scale solar projects are also subject to the California Environmental Quality Act. The California act requires a state agency to prepare an environmental impact report on any project it proposes to carry out or approve that may have a significant effect on the environment. The act generally prohibits the agency from approving a project unless it is modified to mitigate or avoid significant effects on the environment.

BLM Has Received Hundreds of Permit Applications since EPAct 2005 and Authorized 25 Projects

Since 2005, BLM has received hundreds of permit applications for utility-scale renewable energy projects. For wind and solar projects, 17 projects have been authorized, with permit-processing time frames decreasing over time—from about 4 years to about 1.5 years. For geothermal projects, 29 applications were submitted, and construction was approved for 8 projects, with the permit-processing time frames ranging from 1 to 4 years. In all, since EPAct 2005, BLM has authorized projects sited on federal lands with the capacity to generate a total of about 5,450 megawatts of electricity, a substantial increase over the number of megawatts the agencies had authorized before passage of the act, contributing to the act's goal of approving 10,000 megawatts of renewable energy on federal lands by 2015.

Since 2005, BLM Received 416 Wind and Solar Permit Applications and Authorized 17 Projects, and Permitting Time Frames Have Decreased

According to agency officials and responses to our questionnaire, from passage of EPAct 2005 through May 2012, BLM received 416 permit applications for utility-scale wind and solar projects. Over 350 of the applications were for solar energy projects, almost all of which were submitted for development on BLM-managed lands in Arizona, California, and Nevada. Sixty-five applications were submitted to BLM for wind energy projects. One application was submitted for a wind energy project on Forest Service-managed land.[30] Of the 416 applications submitted to BLM, rights-of-way were issued for 20 applications covering 17 projects (7 wind and 10 solar).[31] About 60 percent of the applications were ultimately withdrawn by the applicant or denied by BLM. Table 1 depicts the status of all 416 wind and solar applications. (See also app. I for more information about our methodology for determining how many applications each agency received.)

Table 1. Status of Applications for Utility-Scale Wind and Solar Project Permits on BLM-Managed Lands since EPAct 2005

Status of applications	Wind applications		Solar applications		Total wind and solar applications	
	Number	Percentage	Number	Percentage	Number	Percentage
Authorized	7	11	13[a]	4	20	5
Pending	27	42	82	23	109	26
Withdrawn	17	26	126	36	143	34
Denied	13	20	97	28	110	26
On hold[b]	[c]	[c]	11	3	11	3
Other[d]	1	2	22	6	23	6
Total	**65**	**100**	**351**	**100**	**416**	**100**

Source: GAO analysis of BLM questionnaire results. Because of rounding rules, percentages for some results shown in the table may not add to 100.

[a] A right-of-way was authorized for 13 solar applications; however, these 13 applications covered 10 projects because 1 project had multiple applications.

[b] "On hold" refers to certain solar energy applications for which BLM postponed processing while the solar programmatic environmental impact statement was being completed.

[c] Data not applicable for wind applications.

[d] Applications identified as "other" —for example, those that were closed for administrative reasons other than withdrawal or denial—did not fit in any of the listed categories.

Of the 17 wind and solar projects that had been granted a right-of-way, 7 were in operation and 3 were under construction as of the time of this report. The remainder were delayed for various reasons—such as bankruptcy or changes in the type of technology applicants plan to install—or had been terminated. (See app. II for the status and details of authorized projects for all types of renewable energy.)

Of the 109 wind and solar energy applications that were identified as pending, 2 had been approved through a signed decision at completion of environmental analysis but were not considered authorized because a right-of-way had not been issued, and 20 applications had entered the environmental analysis process as of the time of our review. The remaining applications were in earlier stages of the review process. According to BLM officials, some of these applications are not likely to be considered or approved because of where they stand in the queue of applications for that parcel of land. For example, for some especially desirable parcels, two or more applications may be in the queue for a right-of-way. Applications are processed in the order received, and applications that are further down the queue for the same parcel of land will be considered only if the first application for that parcel is withdrawn or denied. In other cases, BLM officials told us, applicants were no longer actively pursuing their applications but had not yet formally withdrawn them.

Applications were withdrawn or denied for various reasons. According to BLM officials, some applicants withdrew because of concerns regarding the financial market and concerns over the presence of cultural and natural resources located in the project area. Regarding applications that were denied by the agency, BLM officials told us that a majority were denied because applicants did not provide sufficient information—such as technical details associated with their projects—for BLM to process the applications. The next most frequently identified reason for denial was that developers could not demonstrate the technical or financial capability to carry out their projects.

Permitting for the 17 authorized wind and solar projects with applications submitted after EPAct 2005 took from less than half a year to almost 5 years from initial application submission to right-of-way authorization, according to our questionnaire results. Average permitting time frames differed depending on the type of project application; it took about 3.5 years to complete the permitting process for solar projects and about 2.5 years for wind projects.[32] Some of this difference stems from the fact that most wind applications were processed using environmental assessments rather than environmental impact statements, whereas all authorized solar applications were processed using environmental impact statements—and processing applications using environmental assessments took roughly two-thirds as long, on average, as processing applications using environmental impact statements.

In addition, for those wind and solar applications processed using environmental impact statements, the time necessary to complete the application represented about one-third of total processing time. For wind applications processed using environmental assessments, the time to complete the application represented a smaller portion of the total processing time—about 9 percent. Most of the remainder was spent preparing for or conducting the environmental analysis (see figure 2).

BLM officials also reported substantially shorter processing times for wind and solar applications received in more recent years. Specifically, authorized wind and solar project applications submitted in 2006 took an average of about 4 years to process, whereas those applications submitted in 2009 averaged about 1.5 years (see figure 3). According to BLM officials, these time frames may have decreased because BLM field office staff have gained experience in processing renewable energy applications.

BLM Received Applications for 29 Geothermal Projects since 2005 and Approved 8 for Construction, with Permitting Time Frames from 1 to 4 Years

From EPAct 2005 through May 2012, BLM received applications for 29 new utility-scale geothermal projects, most of which were submitted from 2007 through 2009 for development in Nevada. Each component of the application process for developing utility-scale geothermal projects—exploration plans, geothermal drilling permits for individual wells, and utilization plans for constructing and maintaining the geothermal power plant—requires environmental analysis and agency approval. As of the time of this report, of these 29 proposed projects:

- BLM approved 25 exploration plans, with the remaining 4 projects suspended by the applicants.
- Associated with the 25 projects approved for exploration, BLM approved 168 geothermal drilling permits to drill individual wells for developing these projects.

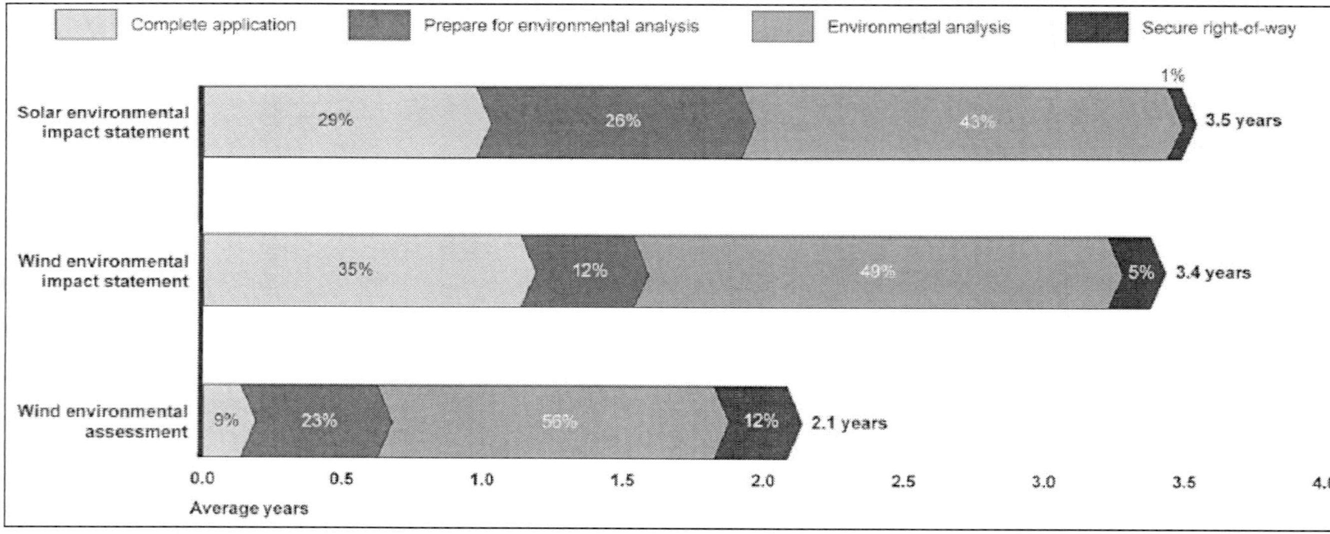

Source: GAO analysis of BLM questionnaire results.

Note: This figure represents 17 wind and solar projects authorized by BLM. As noted in table 1, for one solar project, four applications were submitted, but because these four applications were processed simultaneously, we treated them as a single application in this analysis.

Figure 2. Average Time Frames for Stages in Processing Authorized Solar and Wind Project Applications Submitted after EPAct 2005 Using Environmental Assessments and Environmental Impact Statements.

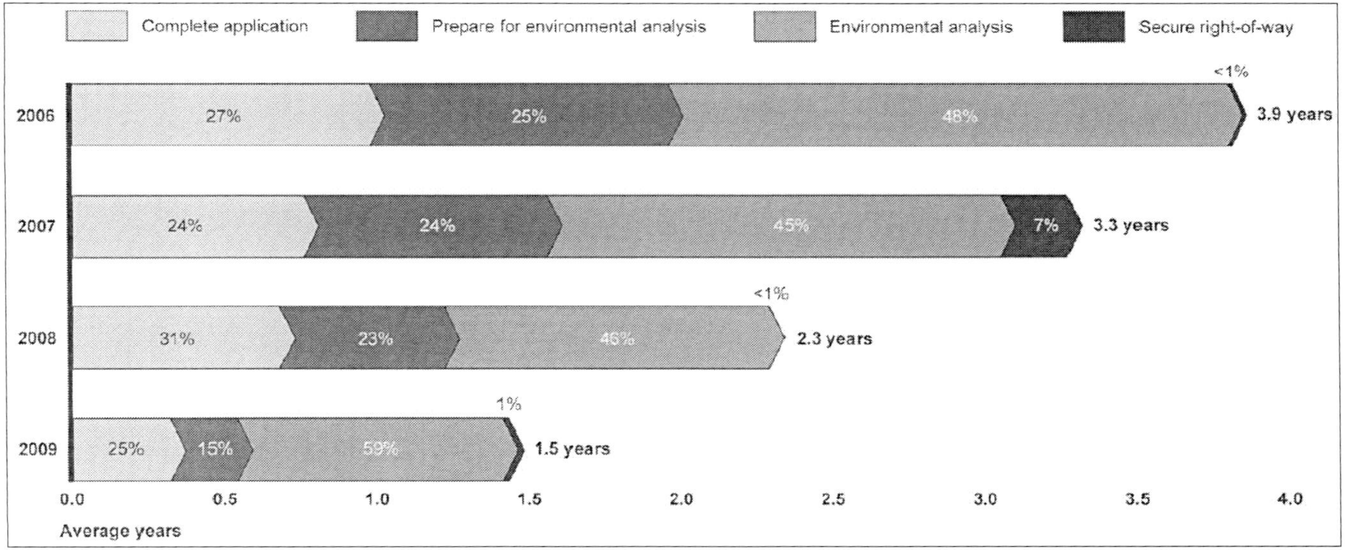

Source: GAO analysis of BLM questionnaire results.

Notes: This figure represents 17 wind and solar projects authorized by BLM. As noted in table 1, for one solar project, four applications were submitted, but because these four applications were processed simultaneously, we treated them as a single application in this analysis. In addition, the data for 2006 represent four applications (one wind environmental assessment and three solar environmental impact statements); for 2007, nine applications (three wind environmental assessments, one wind environmental impact statement, and five solar environmental impact statements); for 2008, two applications (two solar environmental impact statements); and for 2009, two applications (one wind environmental assessment and one wind environmental impact statement).

Figure 3. Average Time Frames for Stages in the Processing of Applications for Authorized Wind and Solar Projects, by Year Submitted.

- BLM subsequently approved 8 of the 25 projects for construction and maintenance; 3 of these projects were operational, and another was under construction.

Some of the 25 applications for which geothermal exploration plans had been approved did not progress to approval of construction because, during the exploration phase, applicants did not find sufficient geothermal resources to justify constructing a plant. In other cases, according to agency officials, financial difficulties constrained applicants' ability to drill wells and construct power plants. A few other projects were early in the exploration process and had not yet moved to the construction phase, according to BLM officials.

Time frames for obtaining the multiple approvals and permits necessary for utility-scale geothermal power plants ranged from 1 to 4 years, in part because of the time needed to explore and find adequate geothermal resources and also because of the time needed to process applications at each approval stage. On average, it took about 1.3 years for approval of exploration plans, about 124 days for approval of drilling permits, and about 1.3 years for approval of construction plans.

Wind, Solar, and Geothermal Projects Applied for and Authorized since EPAct 2005 Total 5,450 Megawatts of Generating Capacity

BLM-authorized renewable energy projects sited on federal lands and applied for since EPAct 2005 have a total electricity-generating capacity of about 5,450 megawatts—a substantial increase over the capacity for renewable energy generation approved before EPAct 2005, which was about 1,360 megawatts. Of this 5,450-megawatt capacity, authorized wind projects contributed about 800 megawatts, solar projects about 4,200 megawatts, and geothermal projects about 450 megawatts.[33]

As previously noted, EPAct 2005 established a goal that the Secretary of the Interior "seek to have approved non-hydropower renewable energy projects located on public lands with a generation capacity of at least 10,000 megawatts of electricity" by 2015. The 5,450 megawatts for authorized projects discussed above contribute to this goal, as do megawatts for projects that have been approved but for which an authorization has not been issued.[34] Other megawatts that contribute to this goal include those from renewable energy projects on non-Interior lands that rely on BLM rights-of-way for "connected actions" essential to the project, such as transmission corridors. BLM officials also told us that the megawatts associated with projects approved before EPAct 2005 contribute toward meeting the goal. In October 2012, with the approval of a large wind energy project, Interior officials announced that the department had surpassed the 10,000-megawatt goal.

AGENCIES HAVE TAKEN SEVERAL STEPS TO FOSTER RENEWABLE ENERGY DEVELOPMENT ON FEDERAL LANDS SINCE EPACT 2005

Since EPAct 2005, the federal land management agencies—primarily BLM but also other Interior agencies and the Forest Service—have developed and revised policies to address renewable energy development on federal lands, formalized collaboration within and across their respective agencies and with state and local governments, and devoted increased resources to process renewable energy permit applications. In addition, BLM—partly in response to our review—has drafted a policy to help guide and assess these efforts to help ensure they are achieving their intended purposes.

Agencies Have Created Policies Aimed at Improving Renewable Energy Development

Since 2005, BLM has developed and revised policies aimed at improving renewable energy permitting and development on its lands through various means. Other agencies—including the Fish and Wildlife Service and the Forest Service—have also developed renewable energy policies. Although some of these steps began before EPAct 2005, BLM officials said they saw the need for continual improvements and changes, given the sharp increase in applications for renewable energy permits since EPAct 2005.[35]

BLM

BLM has developed and revised policies intended for improving renewable energy development on its lands by means of programmatic environmental impact statements, designation of priority projects, instruction memorandums, and rulemaking.

Programmatic Environmental Impact Statements

One of BLM's most comprehensive actions taken with respect to renewable energy was the completion of programmatic environmental impact statements for wind, geothermal, and solar energy development, made final in 2005, 2008, and 2012, respectively (see figure 4). These statements enabled BLM to amend multiple land use plans simultaneously to provide for renewable energy development on BLM- managed lands; in some cases, the statements also established new policies and identified best management practices for energy development. According to BLM officials, the statements were intended to streamline the permitting process for renewable energy development by shortening the amount of time needed for project-by-project environmental impact analyses. The solar statement differed from those for wind and geothermal energy in that it identified specific areas of federal lands most suitable for development—known as solar energy zones—and created incentives, including economic incentives and more streamlined permitting, for development to occur in those locations.[36]

Programmatic environmental impact statements can be particularly useful with regard to renewable energy development because many of BLM's land use plans did not address such development at the utility scale and amending them through a programmatic review was more efficient than doing so one by one. For example, relying on the analysis in the wind statement, BLM amended 52 land use plans, in some cases incorporating wind energy development into land use plans where that energy type had not previously been addressed but where proponents had shown an interest and in other cases restricting wind energy development from wildlife habitat where the agency believed adverse effects could not be mitigated.[37] Without the programmatic statement on wind energy, these land use plans would have had to be amended individually, each with its own associated environmental review.

Moreover, in using such programmatic statements, BLM helped address concerns that NEPA's requirements for environmental analysis can be overly time-consuming. According to BLM officials, the geothermal programmatic environmental impact statement approach has shortened the time it has taken BLM staff to ensure the accuracy of NEPA documentation for site-specific activities such as drilling, in part because it supplied a template for environmental assessments. Additionally, BLM officials anticipate that environmental analysis for applications in the new solar energy zones will be streamlined because BLM can tier these analyses to the relevant programmatic analysis.

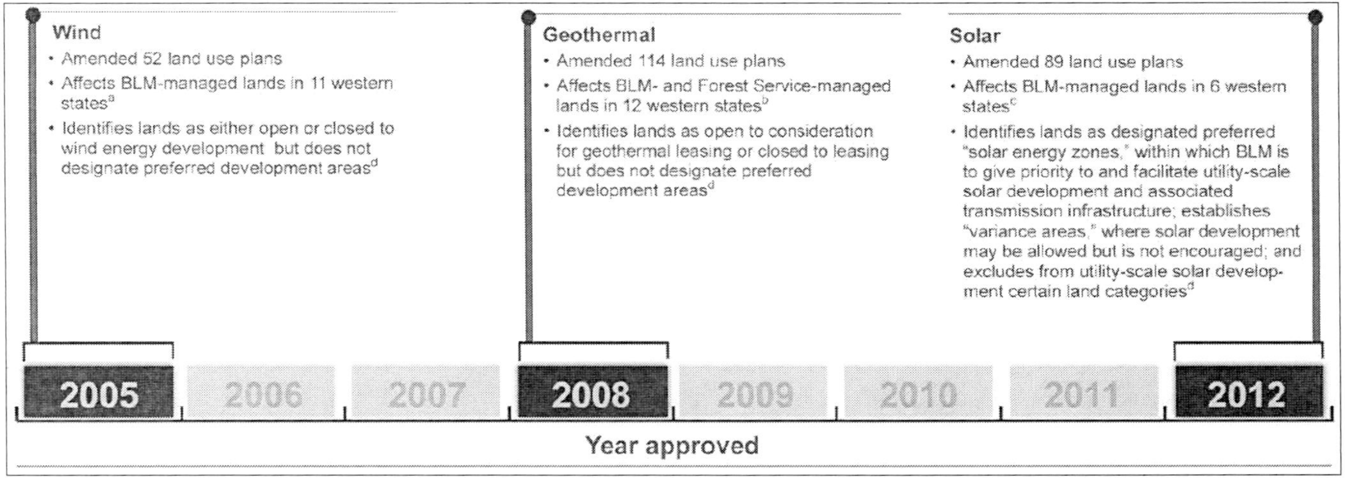

Source: GAO based on BLM documents.

Note: "Year approved" refers to the year when the record of decision was signed for the associated programmatic environmental impact statement. Information in the table generally reflects BLM's actions as described in the relevant record of decision.

[a] Arizona, California, Colorado, Idaho, Montana, Nevada, New Mexico, Oregon, Utah, Washington, and Wyoming.

[b] Alaska, Arizona, California, Colorado, Idaho, Montana, Nevada, New Mexico, Oregon, Utah, Washington, and Wyoming.

[c] Arizona, California, Colorado, Nevada, New Mexico, and Utah.

[d] A variety of areas are closed to development, including, among others, wilderness areas, wilderness study areas, and national monuments.

Figure 4. Summary of BLM's Programmatic Environmental Impact Statements for Wind, Geothermal, and Solar Energy Development.

Priority Projects

In 2009, according to Interior officials, BLM began selecting projects to be given higher priority in application processing so as to focus agency efforts and limited resources on those projects it believed had a greater likelihood of being approved. Recognizing the need for explicit criteria, BLM in 2011 established criteria for prioritizing wind and solar energy projects.[38] These criteria took into consideration natural and cultural resource values—seeking to direct development away from sensitive areas. According to officials, BLM expected that projects on less sensitive lands would take less time to process because these projects would require less consultation, environmental analysis, and mitigation. In addition, BLM also considered those projects for which the application process had progressed far enough to start formal environmental review. Initially, BLM selected priority projects on its own but—recognizing that other agencies are integral to the permitting process or may be affected by project development on BLM-managed lands—it has since coordinated on project selection with other agencies, including the Fish and Wildlife Service, National Park Service, Bureau of Indian Affairs, and Department of Defense.

Some BLM officials told us that additional agency attention paid to priority projects has facilitated processing and streamlined the approval process. For example, the record of decision for all priority projects is now signed by the Secretary of the Interior, which means that the decisions, if challenged, cannot be appealed to the Interior Board of Land Appeals. This board's review can lengthen approval time frames, according to BLM officials.[39]

Instruction Memorandums

Since 2005, BLM has issued multiple instruction memorandums to its field offices containing new policies aimed primarily at development of wind and solar energy. A significant increase in wind and solar permit applications beginning in 2007—many of which BLM officials said were speculative in nature or not detailed enough for consideration—coupled with BLM's relative inexperience in processing such applications made managing the applications a challenge. Several of the memorandums stemmed from a lessons-learned workshop on renewable energy held in January 2011, where officials from several Interior agencies, as well as industry groups, identified challenges and developed suggestions for improvement. The following month, on February 7, 2011, BLM issued three instruction memorandums specifically targeted at clarifying NEPA documentation, facilitating the application review and approval process, and improving the quality of project applications. The instruction memorandums are as follows:

- *A memorandum aimed at clarifying BLM's policy under NEPA as it relates to analyzing applications for utility-scale renewable energy projects.*[40] This memorandum provided guidance to BLM staff on developing and presenting certain information required for environmental impact statements related to renewable energy projects. For example, as part of the NEPA process, BLM is to analyze reasonable alternatives for development. Given the various technologies available for renewable energy development, some alternatives may be proposed that are not technically or economically feasible. This memorandum states that reasonable alternatives include "those that are practical or feasible from the technical and economic standpoint and using common sense, rather than simply desirable from the standpoint of the applicant." It also notes that information about the applicant's interest and objectives, including any constraints or flexibility in a proposal, is needed to help BLM determine which alternatives to analyze in detail under NEPA and, conversely, which alternatives to eliminate from detailed analysis.
- *A memorandum providing updated guidance to facilitate the application review and approval process by directing early coordination between BLM and other stakeholders.*[41] This memorandum requires that all prospective applicants participate in at least two preapplication meetings with BLM before the agency can accept an application for solar or wind projects.[42] The first such meeting, between an applicant and BLM, helps ensure that the applicant is familiar with BLM's right-of-way process and allows discussion of issues such as the applicant's proposal for the project, any land use and siting constraints, potential environmental issues, and potential alternative site locations. The second meeting is to initiate coordination with other federal agencies, such as the Fish and Wildlife Service, as well as with tribal, state, and local government agencies; to provide an additional opportunity to discuss potential environmental and siting constraints; and to modify the proposed project, if necessary, before an application is submitted.
- *A memorandum aimed at improving the quality of project applications, thereby enabling BLM officials to better identify*

applicants with a serious interest in project development.[43] This memorandum sought to enable BLM to distinguish between applicants that are serious about developing a renewable energy project and those considered to be land speculators. To enable these distinctions, BLM clarified the expectation that wind and solar project applications are to include detailed project descriptions before BLM can begin further processing and that the agency will not accept plans that are still in the conceptual phase. Because right-of-way permit applications are processed in the order in which they are received, identifying serious applicants early in the process can help ensure that BLM's time is spent reviewing and processing projects likely to come to fruition.

Rulemaking

BLM in 2011 undertook two rulemaking actions to facilitate the development of utility-scale wind and solar power on federal lands. First, in April 2011, BLM issued a temporary rule to immediately prevent the filing of mining claims in areas contemplated for wind or solar energy development, plus a related rule that would make permanent the temporary ban on mining claims. According to agency statements in the temporary rule, over the previous 2 years, hundreds of new mining claims were filed for areas where wind and solar right-of-way energy applications had been submitted. Also, according to these agency statements, many of these claims were likely to be speculative. Specifically, the temporary rule stated that these claims were filed not for true mining purposes but rather for the mining claimant to try to compel some kind of payment from the renewable energy applicant before relinquishing the mining claim.[44] BLM officials told us that the agency expected to publish the final rule in January 2013. Second, in December 2011, BLM solicited public comments to be used in preparing a proposed rule to establish a competitive process for leasing federal lands for solar and wind energy development.[45] According to agency statements in the proposed rule, a competitive process—rather than the first-come, first-served process currently in use—would enhance the agency's ability to capture fair market value and ensure fair access to leasing opportunities.[46] The rulemaking would establish competitive bidding procedures for lands within designated solar and wind leasing areas,[47] define qualifications for potential bidders, and structure the financial arrangements necessary for the process. The agency expects to issue a proposed rule in January 2013.

In 2007, BLM issued a rule revising the agency's geothermal resources leasing regulations to implement EPAct 2005. This rule established, among other things, a new process for competitive leasing, under which BLM is generally to issue a lease to the highest bidder.

Other Agencies

In addition to new policies created by BLM, the Fish and Wildlife Service in March 2012 issued wind energy guidelines aimed at improving utility- scale wind energy development to reduce potential impacts to species of concern—including migratory birds, bats, bald and golden eagles, and sage grouse—regardless of whether projects are proposed for federal or private lands.[48] Among other aims, these guidelines are intended to promote compliance with wildlife laws and regulations, encourage scientifically rigorous assessments proportionate to risks facing species of concern, and mitigate potential adverse effects on species of concern and their habitats. The guidelines assist developers in identifying species of concern that their proposed projects may affect and also discuss risks to those species. Adherence to these guidelines is voluntary. Regarding Forest Service-managed lands, in August 2011, the Forest Service amended its special use directives to add provisions specific to wind energy projects.[49] Previously, agencywide policy was to deny wind energy development proposals if the proposed development could reasonably be accommodated on lands not managed by the Forest Service. Now, however, this policy is to be considered in conjunction with the agency's encouragement of wind energy facilities on Forest Service-managed lands to help meet the nation's energy needs. No such directives exist for solar projects, although agency officials told us they are drafting such provisions and expect to complete them in 2013. In general, officials with the Forest Service told us that interest in utility-scale wind and solar development on their lands is considerably less than for BLM-managed lands and that, as a result, the Forest Service has relied primarily on BLM guidance when considering proposals for development.

Agencies Have Taken Steps Intended to Improve Coordination within and across Agencies

Since EPAct 2005, the federal land management agencies have taken steps intended to improve coordination as a way to streamline the permitting process and promote renewable energy development on federal lands in general. Specifically, coordination has been formalized through regularly established meetings and memorandums of understanding across Interior agencies, as well as with other federal agencies and state and local governments.

To facilitate coordination on renewable energy activities among its component agencies, Interior instituted weekly meetings among its component agency officials at the national level to discuss issues concerning individual renewable energy applications and projects, particularly those identified as priority projects. These "strike team" meetings were intended to formalize coordination to ensure that officials across component agencies are aware of concerns that could affect the development of projects—including critical habitat, cultural or tribal issues, conflicts with national park boundaries or interests, and other environmental issues—and have a recurring forum where such concerns may be aired and resolved. BLM's national and state renewable energy coordination offices also hold weekly meetings to discuss the status of projects and other issues as they come up.

In several cases, coordination between Interior agencies within certain regions has been formalized through a memorandum of understanding. In January 2008, for example, in recognition of the increase in renewable energy permit applications for projects in Southern California, BLM's California Desert District and the Fish and Wildlife Service's Ventura and Carlsbad offices signed a memorandum of understanding to help ensure efficient completion of required consultation under the Endangered Species Act. This memorandum defines the process, products, actions, time frames, and expectations needed to complete the process. In addition, in June 2011, a memorandum of understanding was signed between BLM's California state office and the National Park Service's Pacific West Region. The memorandum generally documents coordination procedures the two agencies are to follow in instances where proposed renewable energy projects in the region may affect lands under the jurisdiction of the National Park Service. According to Park Service officials, a similar memorandum is being explored between the National Park Service's Pacific West Region and BLM's Nevada state office. Interior officials told us that, given the success of the California effort, the agency is discussing potential amendments to agency manuals to ensure that closer coordination occurs between each BLM state office and relevant National Park Service regional offices. Before putting such amendments in place, however, the agency wants to evaluate different tools to encourage and reward staff for engaging in coordination.

BLM has also formalized its coordination on renewable energy with other federal agencies, as well as with state and local entities. For example, in 2006 BLM and the Forest Service signed a memorandum of understanding to implement certain geothermal leasing and permitting provisions of EPAct 2005, in part to reduce the backlog of geothermal leasing applications. In addition, over the past decade, BLM and the Department of Energy's National Renewable Energy Laboratory have maintained a relationship whereby the laboratory has provided expertise to BLM in support of BLM's management of renewable energy development. According to laboratory officials, the relationship has evolved so that the laboratory provides BLM with assistance on a systematic rather than task-by-task basis. For example, in 2011 BLM and the laboratory signed an interagency agreement for the laboratory to provide technical assistance and training for fiscal years 2011 through 2013. Given the laboratory's expertise, laboratory officials have in some cases assisted BLM offices with individual renewable energy applications by evaluating the technical feasibility of certain proposals. In addition, BLM state offices are working in partnership with state and local agencies. In California, multiple federal land management agencies are working with the California Energy Commission and the California Department of Fish and Game in a collaborative effort to develop the Desert Renewable Energy Conservation Plan for 22.5 million acres of public and private land in Southern California. This effort, initiated under a 2008 memorandum of understanding, is intended to develop a conservation strategy to provide for protection and conservation of the natural resources within the Mojave and Colorado Desert Regions while allowing solar and other renewable energy development in a manner that avoids or minimizes environmental impacts. According to Interior officials, other BLM efforts to coordinate with states have occurred in Wyoming and, to a lesser degree, in Arizona, Nevada, and New Mexico. Officials told us that success in developing renewable energy projects depends on collaboration with states, making such partnerships important.

Agencies Have Devoted Additional Resources to Renewable Energy Development

From fiscal year 2010 to fiscal year 2012, BLM reported more than doubling the program funding devoted to wind and solar energy activities, from about $8 million to about $16.5 million.[50] During that time, BLM added

64 full-time-equivalent staff—tripling its staff from 32 in fiscal year 2010 to 96 in fiscal year 2012. In addition to these funds, BLM uses cost- recovery fees, which it is authorized to recover from applicants to pay for processing applications for wind and solar energy development.[51] From fiscal year 2009 to fiscal year 2012, BLM reported collecting about $16 million through these fees;[52] BLM officials estimated that cost-recovery fees represent about half of the total funds received by BLM field offices.

In contrast, funding and staffing for geothermal energy leasing and permitting declined in recent years after the expiration of temporary funding authority created by EPAct 2005. Specifically, the act established the Geothermal Steam Act Implementation Fund to be used to expedite the development of geothermal energy.[53] The act required that certain rents and royalties paid as part of geothermal leases go into the fund until the end of fiscal year 2010. BLM used these funds for coordinating and processing geothermal, permits, among other uses. Authorization for this fund has expired, and after expiration, BLM's geothermal-related obligations declined from about $7.9 million to about $2.7 million from fiscal year 2010 to fiscal year 2012, and the number of full-time- equivalent staff was reduced from 45 to 35 during that time.

The funding and staffing increases for wind and solar energy development were in large part used by BLM to establish renewable energy coordination offices at the national and state levels to, among other duties, help process wind and solar right-of-way permit applications in the four western states with significant renewable energy activity: Arizona, California, Nevada, and Wyoming. BLM also established similar but less well-developed entities, known as renewable energy teams, in several of the other western states with less-significant renewable energy activity.

Interior agencies besides BLM have also recently devoted additional resources toward renewable energy efforts. To accommodate the workload associated with increased interest in renewable energy development, the National Park Service in 2010 hired a permanent, full- time national external renewable energy coordinator and six full-time- equivalent staff focused on coordination for renewable energy permitting. According to agency officials, these positions are supported by using funds from the agency's existing programs. Other National Park Service staff also contribute time to renewable energy efforts on BLM-managed lands as a collateral duty. For example, agency officials told us that several agency staff representing national parks in

Southern California have spent time working with BLM and project applicants on renewable energy proposals near national parks in the region; these officials told us they work with BLM and applicants to minimize potential negative effects on park resources, such as impaired views, increased noise and light pollution, and disruption to area wildlife.

The Fish and Wildlife Service also reported receiving more appropriated funds for engagement in renewable energy permitting. Specifically, in fiscal year 2010, the agency's Conservation Planning Assistance Program received a $1.5 million increase for technical assistance on renewable energy projects. In fiscal year 2011, this program received an additional $2 million for these activities, for a total increase of $3.5 million. This $3.5 million increase continued for fiscal year 2012. The program's funding covered not only time spent contributing assistance to projects proposed on federal lands but also to those proposed on private lands. Some of this funding helped address region-specific increases in workloads. For example, according to agency officials, in fiscal years 2011 and 2012, the Fish and Wildlife Service's Pacific Southwest Regional Office was able to add two to three full-time-equivalent staff positions at the state level and to conduct consultation and project planning at five offices working on renewable energy in the region. Nonetheless, several Fish and Wildlife Service officials told us that they completed their increased federal lands renewable energy-related work as a collateral duty done in conjunction with other work responsibilities.

Interior has taken steps to address its component agencies' concerns over the amount of time and funding they commit to the renewable energy permitting process, according to agency officials. For example, Interior has prepared a draft secretarial order delegating authority for cost recovery to the Bureau of Indian Affairs, Fish and Wildlife Service, National Park Service, and others to allow these agencies to collect cost- recovery fees to help pay for their activities as part of the permitting process. A final order is expected to be issued in 2013.

According to Forest Service officials, interest and activity in developing wind and solar energy projects on Forest Service-managed lands have been at a relatively low level, and no specific Forest Service increases in funding or staff comparable to BLM's have been dedicated to renewable energy development. Forest Service officials told us that funding is received through recovery of costs from applicants when needed for specific renewable energy activities. The agency maintains two expert advisors on technical and legal issues related to geothermal leasing, as well as other personnel who work on renewable energy projects and policy as collateral duties. Nevertheless, the Forest Service does not dedicate specific funding to wind, solar, or geothermal energy development; rather, funding for such development is included in agency budget line items covering broader activities. According to Forest Service officials, they do not track the amount the agency spends on renewable energy permitting activities.

BLM Plans to Review Its Renewable Energy Activities

To help ensure the efficiency and effectiveness of its renewable energy activities, BLM issued a new instruction memorandum in December 2012 aimed at providing BLM offices involved in renewable energy activities with a better understanding of renewable energy policies and regulations and to provide clarity and consistency in the goals of the agency's renewable energy activities. The memorandum—developed partly in response to our review—directs the national renewable energy coordination office to establish an Oversight and Implementation Plan. This plan is to establish an internal review to be performed annually by the national renewable energy coordination office, beginning in fiscal year 2013. The review is intended to ensure BLM compliance with renewable energy regulations and policies and to ensure that guidance is applied appropriately and consistently throughout BLM. In addition, the plan is to call for collaboration among the various offices in the development of future renewable energy policies.

During our review, some agency officials and stakeholders acknowledged the importance of greater resources and coordination in facilitating the permitting process but also informed us of areas for improvement— thereby indicating the importance of assessing the permitting process. For example, one agency official and some stakeholders told us that BLM field offices were not always consistent in their approach to renewable energy permitting or in their willingness to coordinate on renewable energy projects; they said that BLM could do more to help ensure consistency in coordination. An ongoing oversight and improvement plan can assist BLM in identifying needed improvements to its renewable energy permitting process.

Factors Related to Permitting, as Well as Market and Other Forces, Affect the Pace of Renewable Energy Development on Federal Lands

According to BLM officials' responses to our questionnaire and interviews we conducted with others, including industry representatives, many factors can affect the pace of renewable energy development on federal lands. Some of these factors relate to the land management agencies' approval and permitting processes, primarily BLM's, with BLM respondents and others identifying some factors that facilitate and others that hinder the permitting process. Other factors identified as important include some related to broader market forces, which operate outside of the permitting process and therefore beyond the agencies' direct control.

Agency Officials and Others Identified a Number of Factors Related to Permitting That Can Facilitate or Hinder the Process

Agency officials, industry representatives, and other stakeholders we spoke with identified several key factors affecting the processing of renewable energy applications. Factors facilitating the permitting process include coordination among the involved parties and resources the agency can devote to permitting; factors that may hinder or slow permitting include the quality of submitted applications and managing for the presence of natural and cultural resources on proposed development locations.[54]

Facilitating Factors

Factors identified as facilitating the permitting process are generally related to coordination among the parties involved in permitting and the availability of BLM resources.

Coordination among Parties Involved in the Permitting Process

BLM respondents identified the quality of coordination between parties involved in permitting individual projects as among the top factors that facilitated the permitting process for each of the three energy types we reviewed. Such coordination occurs throughout the application process, including the initial request for an application; when key portions of the application, such as the plan of development, are made final; when environmental analyses are conducted and NEPA documentation is prepared; and the issuance of the right-of-way. It also happens among several parties—between the applicant and BLM staff, among staff within BLM, and between BLM staff and staff from other federal and nonfederal agencies.

BLM respondents identified one forum for coordination as particularly helpful in facilitating the permitting process: participation in preapplication meetings, where coordination occurs among multiple parties. These meetings, which take place before an applicant submits an initial application to BLM, are meant to provide opportunities for applicants, BLM officials, and other stakeholders to discuss the permitting process, early concerns about project proposals, and other issues. According to some respondents to our questionnaire, applicants who spent time coordinating with stakeholders early could generally move through the permitting process more easily than those who had not. Several respondents acknowledged that preapplication meetings could be time- consuming, but these respondents also said that such meetings could help ensure that applicants understand BLM's permitting process and required documentation, select appropriate locations to site projects (e.g., to avoid areas with environmental or other constraints), and submit an adequate application. Concerning one application, a respondent explained that frequent and early coordination between BLM and the applicant clarified expectations for both parties as to what information the applicant needed to provide to BLM, thus preventing delays associated with unanticipated surveys or studies that might otherwise have been required.

Such coordination often involved other federal agencies, which BLM respondents and others identified as important. Some Fish and Wildlife Service officials told us that meeting with applicants at the preapplication stage can help ensure that potential harm to species and ways to mitigate that harm are identified and addressed in the application. For example, particularly with wind energy projects, officials said they work with applicants to mitigate not only harm to threatened and endangered species and their habitats under the Endangered Species Act but also harm to other species protected under laws such as the Migratory Bird Treaty Act and Bald and Golden Eagle Protection Act.[55] Fish and Wildlife Service officials also said that they work with applicants at the preapplication stage to prepare species conservation plans and issue species-specific Fish and Wildlife Service permits. These officials said that these early meetings and documentation from applicants provide both the applicant and the agencies greater certainty that the applicant can comply with relevant species protection laws. Similarly, a respondent told us that for one approved project, early coordination between the applicant and officials from a nearby Air Force base helped the applicant select a site that was less likely to interfere with the base's training needs, thereby limiting the potential for conflict as the application progressed through the permitting process. BLM respondents also identified coordination with state, local, and tribal governments as facilitative, although to a lesser degree than coordination with federal agencies.

Most representatives of industry and environmental groups we spoke with commended BLM's efforts to enhance coordination, but they suggested that improvements could still be made. For example, one industry representative suggested that preapplication meetings could be more beneficial if BLM designated one staff member as coordination leader for a particular project. This individual could then be responsible for setting deadlines and holding the various other participants accountable for completing their respective tasks in a timely manner.

Availability of BLM Resources to Support Permitting

BLM respondents frequently cited the availability of resources as an important factor in the permitting process. They identified the importance of (1) cost-recovery funds provided by applicants to cover agency permitting costs for wind and solar energy projects and (2) internal BLM resources used to support activities such as administration and policy development.

The cost-recovery funds that BLM requires applicants to submit before it processes wind or solar applications pay for agency permitting expenses, such as reviewing applicant-provided information and studies, preparing environmental analyses, and recording information in agencywide databases. According to BLM officials, BLM's expenses are generally substantial given the scale of most renewable energy projects and required reviews, ranging from $50,000 to almost $400,000. Without receiving at least a portion of its expenses in the form of cost-recovery funds, BLM does not begin processing an application; any delay in receipt of these funds can cause additional delays throughout the permitting process. Several respondents noted that in certain cases the applicants never provided cost-recovery funds, even after multiple BLM requests.

In addition to cost-recovery funds, BLM provides internal resources to support administrative activities not specific to individual applications, as well as to maintain adequate staffing with sufficient expertise to process renewable energy applications. BLM officials and industry representatives told us that when BLM first began receiving applications for utility-scale solar development, it did not have expertise in either solar energy development or in processing permits for this type of right-of-way. Over the last few years, however, BLM has increased the number of its staff and the amount of funding dedicated to renewable energy development and has provided training opportunities to its staff who process applications; agency officials and industry representatives also told us that staff have become more knowledgeable through experience.

Representatives from industry and environmental groups and officials from agencies other than BLM concurred that BLM staff have in more recent years become more knowledgeable about renewable energy in general and the permitting process specifically. Several commented that the pace at which an application was processed depended on the abilities of the assigned project manager and the emphasis placed on renewable energy by the field office processing the application.

Other Factors

BLM respondents to our questionnaire commonly identified two other factors as facilitating the permitting process, although these factors were cited less frequently than those related to coordination or resources. One of these other factors was agency policy and guidance related to energy development. Specific policies identified by respondents as particularly helpful for wind and solar applications included programmatic environmental impact statements; instruction memorandums from BLM, including the memorandum addressing NEPA compliance for utility-scale renewable energy development; and a 2003 memorandum containing interim guidance from the Fish and Wildlife Service on avoiding and minimizing wildlife impacts from wind turbines.[56] In addition, one industry representative told us that BLM's solar programmatic environmental impact statement is effective in screening lands for the presence of cultural and biological resources. Another industry representative called the

Fish and Wildlife Service's 2012 wind energy guidelines an effective tool that clearly describes agency expectations.

Another facilitating factor identified was the selection of applications as priority projects. Several respondents and industry representatives told us that a priority designation helped because the various stakeholders involved in the permitting process dedicated their attention to reviewing priority projects. Others, however, were less certain of benefits. A few industry representatives told us they did not observe a reduction in application processing times. Also, one BLM respondent told us that, although the designation of a solar application as a priority project facilitated the permitting process for this project, it also encouraged officials to process the application faster than was appropriate, given that the necessary biological and cultural surveys had not been completed.

Hindering Factors

BLM respondents to our questionnaire and other agency and industry representatives we spoke with identified several key factors that could delay the processing of renewable energy applications, such as application quality and other applicant-related factors, as well as factors related to managing for natural and cultural resources.

Respondents identified the quality of a renewable energy application, including key components such as the plan of development for wind or solar projects, as among the top factors hindering the permitting process. The plan of development for wind and solar projects is to include information on structures and facilities associated with the proposed project, engineering specifications, maps, and other items. Similarly, applications for geothermal power plants are to include a detailed utilization plan, completed and signed facility construction permit, and completed and signed site license. According to BLM officials, this information allows them to determine the scale and scope of the proposed development and begin analyses to determine project feasibility. The agency does not begin processing applications until the information is complete. Since many applications require more information than is initially submitted, this step may delay processing, depending on how much additional information is requested by BLM and the applicant's ability to respond. For example, one respondent to our questionnaire said that in addition to poorly defining the project, an applicant continually altered key application components—including project scope, design, and technology—which made it difficult for BLM officials to analyze impacts on natural and cultural resources and process the application in a timely fashion. BLM officials also told us that some applicants did not include detailed information necessary to begin review, such as technical plans for engineering and hydrological design or stormwater management, which delayed processing of the applications. When critical components like these are missing from an application, delays result. BLM questionnaire respondents reported that in some instances applicants did not respond to the agency's requests for additional information or responded with incomplete information, leading to denial of their applications. As noted, our analysis showed that the most prevalent reason for denial was that an applicant did not provide information BLM requested that was needed to process the application.

On the other hand, several industry representatives told us that the amount of documentation necessary to develop a project on federal land is, in their view, at times excessive. Some representatives told us that additional information and studies requested by the agency can be time- consuming and expensive to provide, thereby delaying projects, and, in their opinion, are not always necessary. For example, one applicant for a wind energy project wanted to incorporate into the application studies on golden eagles, a protected species, that had recently been completed for areas close to the proposed project site. The Fish and Wildlife Service, however, requested site-specific studies, which could take 3 to 5 years.

Similarly, several questionnaire respondents identified applicants' requests for BLM to delay the processing of their applications as hindering the permitting process. The respondents noted several reasons that applicants might make such requests, including the need for more time to respond to BLM's requests for additional information and changes to agency policies or fees. For example, applicants might have requested a delay because they wanted to wait until the effects of BLM's recently issued programmatic environmental impact statement for solar energy development could be determined. In other cases, costs associated with development—such as BLM rental fees paid by the applicant for siting a project on federal land or the costs required for mitigation efforts necessary at the project site—can cause an applicant to request a processing delay. One questionnaire respondent reported that an applicant for a solar energy project requested a BLM delay when BLM quoted rental rates higher than the applicant expected.

BLM questionnaire respondents identified several other factors that, to a lesser degree, hindered the processing of applications, including managing for the presence of threatened or endangered species, or other species of concern, and managing for the presence of tribal, cultural, or historic resources. For example, for one pending solar application in Nevada, BLM officials noted that the applicant is performing additional wildlife and environmental analyses and consultation with the Fish and Wildlife Service to avoid impacts to the desert tortoise. For other projects, managing for the presence of tribal, cultural, or historical resources can contribute to the time it takes to process an application. In some areas, a burial site or a place of sacred significance may require extensive tribal consultation. For one project, an industry representative told us that extensive tribal consultations led to the applicant's changing the project's design about 15 times. Ultimately, according to this representative, the applicant spent thousands of hours on archaeological reviews and moved the project to prevent any degradation of culturally sensitive areas. This representative told us that the lack of specificity in the regulation guiding tribal consultation makes it difficult to ascertain whether an applicant's proposed mitigations for sensitive resources are likely to be adequate. As a result, according to this representative, lawsuits may still ensue, further delaying development even after a project has been approved.

BLM questionnaire respondents also identified requirements for state or local laws as factors that can hinder the permitting process. Specifically, several respondents told us that compliance with the California Environmental Quality Act can add procedural layers to processing applications in California.

Market and Other Forces Outside the Permitting Process Also Play a Role

BLM questionnaire respondents, other BLM and federal agency officials, and industry representatives we spoke with also identified as important some factors outside of the permitting process and beyond agency control. Specifically, market forces and other factors influence whether renewable energy projects on federal lands are proposed, funded, and completed. Some of these factors relate to demand for renewable energy, including competition from electricity generated through other sources, and others relate to supply, including the availability of electrical transmission lines. Such forces can play a key role for any utility-scale renewable energy project by setting the financial backdrop for the project, according to industry representatives. For an applicant to secure financing for renewable energy projects on federal lands, investors generally require some assurance that the costs to construct the project and produce the energy will not exceed revenues from selling that energy. For example, BLM officials we spoke with told us that one reason that some renewable energy projects were withdrawn was that financing was not available because of concerns about whether the project could repay its investment costs. These officials said that financing for permitting, construction, and other aspects of development is more likely to be available to applicants who demonstrate that demand is sufficient and that the cost of supplying power allows for profitability.

According to agency officials and industry representatives, demand for electricity from renewable sources can be influenced by such factors as competition from other sources of energy and the presence of renewable portfolio standards. As natural gas prices have decreased in recent years,[57] renewable energy sources may have become less attractive to electricity purchasers and investors. Because wind and solar energy development provides an intermittent supply of electricity, purchasers and investors may find other, more steady supplies of electricity more attractive. In addition, demand for renewable energy may be affected by the presence of renewable energy portfolio standards;[58] at least 29 states have such standards in place, under which utilities are required to derive a minimum percentage of total electricity they sell from renewable energy sources. Such standards may increase demand for renewable energy overall, but because an increasing number of these state standards have been or are nearly met, the utilities required to comply with these standards may no longer be interested in purchasing additional renewable energy within the next several years, according to industry representatives. BLM respondents noted that applicants sometimes struggled to find buyers for their electricity and, consequently, had difficulty pursuing their projects. Industry representatives told us that establishing power purchase agreements—contracts between energy producers and energy purchasers—was a key part of the success of their renewable projects, and several questionnaire respondents noted that some wind and solar projects could not move forward because the applicants were unable to secure such agreements. In one case, a BLM official noted that the inability of a solar project to secure a power purchase agreement contributed to the cancellation of a 700-megawatt project after BLM had authorized the project's right-of-way permit.

Officials we spoke with also identified factors influencing the cost of supplying renewable energy—most notably the accessibility of transmission lines and the availability of government incentives—as having a role in the pace of energy development on federal land. According to agency officials and industry representatives, the availability of nearby power transmission and distribution lines and access to these lines are critical to the economic viability of a renewable energy project, regardless of whether the project is located on federal or nonfederal lands. Often, however, renewable energy sources are abundant in areas where transmission lines are scarce—increasing the overall difficulty and cost of supplying renewable energy. Moreover, the construction of new transmission lines can be costly and can face its own regulatory and environmental challenges, including separate environmental analyses. For example, several BLM officials in Wyoming told us that not having a sufficient transmission infrastructure has been a deciding factor in some companies' decision to suspend or halt their renewable energy projects. In addition, we were told that the availability of government incentives can affect the cost of producing electricity from renewable resources. For example, the federal tax code includes special tax incentives for solar energy development, which can help increase the potential profitability of planned solar projects. Such tax credits can be useful, although less so if available financing is limited, because without financing for an initial investment, projects that might benefit from tax credits would not be built in the first place. As part of a federal response to the 2008 recession, the American Recovery and Reinvestment Act of 2009 expanded a number of existing tax incentives, including ones for renewable energy development. Several industry and government officials we spoke with cited the Energy Production Credit (also known as the Production Tax Credit) for wind facilities in particular as an important incentive for encouraging the development of several new wind projects. However, this tax credit was scheduled to expire at the end of calendar year 2012,[59] which, according to some industry representatives, likely contributed to a reduction in the number of proposed wind projects. According to a respondent to our questionnaire, this uncertainty hampered one applicant's ability to attract investors, and the applicant ultimately found the project no longer economically viable.

AGENCY COMMENTS AND OUR EVALUATION

We provided a draft of this report for review and comment to the Departments of Agriculture, Energy, and the Interior. In written comments, the Department of Agriculture concurred with our findings, while the Departments of Energy and the Interior had no comments.

Sincerely yours,

Anne-Marie Fennell
Director, Natural Resources and Environment

APPENDIX I. OBJECTIVES, SCOPE, AND METHODOLOGY

This report examines (1) the status of renewable energy permitting on federal land, including time frames for processing permits applied for since the Energy Policy Act of 2005 (EPAct 2005); (2) actions the agencies have taken to facilitate renewable energy development on federal lands, particularly since the passage of EPAct 2005; and (3) factors affecting renewable energy development on federal land.

To address our first objective, we identified applications for utility-scale onshore wind and solar projects submitted to the Bureau of Land Management (BLM) after enactment of EPAct on August 8, 2005, through May 31, 2012, by asking the BLM renewable energy coordination office to query the BLM database designed to track such applications. We also identified geothermal drilling permit applications and associated utility- scale geothermal project applications submitted to BLM during the same time period by asking BLM to query the agency database designed to track geothermal permits. We contacted BLM because it is the agency responsible for permitting almost all applications submitted for renewable energy development on federal lands. We used the results of this query to group the permits according to the utility-scale projects of which they were a part. After reviewing the lists of applications and correcting for any obvious duplication and other errors, we sent each BLM state office a list of applications for projects located in that state and asked officials to verify our lists' accuracy. In some instances, we deleted some applications because BLM officials informed us that certain applications were submitted to BLM before enactment of EPAct or were submitted for resource exploration or testing without being part of utility-scale development. BLM officials also identified applications that met our criteria but did not appear in their database; we added those applications to our analysis. We identified a total of 65 permit applications for wind projects, 351 for solar projects, and 29 for geothermal power plants, as well as 405 applications submitted for geothermal drilling permits from August 8, 2005 through May 31, 2012.

We distributed an electronic questionnaire to BLM officials for all projects meeting our criteria. We developed questionnaires for each energy type. For each application, we asked respondents to identify the dates of certain milestones (e.g., the date the application was received or the date it was considered complete) and the factors that facilitated or hindered processing of that particular application. Although the databases contained some of the milestone information we collected through our questionnaires, they did not include information for each of the dates needed for purposes of our analysis. To encourage questionnaire recipients to respond, we held meetings with national and state BLM officials to explain the questionnaire's purpose, followed up by telephone and e-mail, and obtained a 100 percent response rate for projects meeting our criteria.

We sent the completed questionnaires to a third-party contractor to compile the results, checked the compiled results to ensure their accuracy and reliability, and followed up as needed to clarify incomplete or ambiguous responses. In some cases, we eliminated certain responses from our analysis because we could not obtain sufficiently complete information from the respondent. Because our questionnaire did not sample from a population, no sampling errors occurred. Nevertheless, the results of any questionnaire may be subject to errors, commonly referred to as nonsampling errors. For example, differences in how a particular question is interpreted, in the sources of information available to different respondents, or in how data are entered into a database or analyzed can introduce unwanted variability into questionnaire results. We took steps in questionnaire development, data collection, and data analysis to minimize these nonsampling errors. For example, before developing the questionnaires, we met with BLM officials at headquarters, state, and field offices to discuss the permitting process. We also reviewed current policies and legislation relevant to our questions and the analysis of the responses. The questionnaire was designed by GAO questionnaire specialists in conjunction with staff having subject-matter expertise. We pretested a draft of the questionnaire with officials from five BLM state offices—Arizona, California, Nevada, Utah, and Wyoming, the states that received the greatest number of applications—to ensure that the questions were relevant, clearly stated, and easy to understand.

We also conducted semistructured interviews with the special-use coordinators in each of the nine regional offices of the Forest Service. During these interviews, we requested information on the number of applications received and time frames for processing applications, as well as on factors facilitating and hindering the permitting process.

To address our second objective, we reviewed relevant laws, regulations, and agency policies and guidance. We also interviewed officials from the Department of the Interior and its four land management agencies—BLM, the Bureau of Indian Affairs, the Fish and Wildlife Service, and the National Park Service—as well as from the Department of Agriculture's Forest Service. In addition to officials in these agencies' headquarters, we interviewed BLM officials from California, Colorado, Nevada, and Wyoming. We selected California, Nevada, and Wyoming primarily because they represent areas where substantial renewable energy development on federal lands is taking place, and substantial agency resources are devoted to processing renewable energy applications. We selected Colorado primarily to obtain the perspective about the program from a state with less activity and fewer resources. We interviewed Fish and Wildlife Service officials from the agency's Mountain-Prairie and Pacific Southwest Regions, both selected because of substantial renewable energy development occurring in the regions, and from one state office within each of these regions. We interviewed a National Park Service official from a national park in California because of his experience with renewable energy development near the park. In addition, we interviewed Forest Service officials from all nine Forest Service regions and officials in the wind, solar, and geothermal research programs at the Department of Energy's National Renewable Energy Laboratory in Golden, Colorado. We reviewed strategic plans, agency reports, and reviews of renewable energy development on federal lands; memorandums of understanding between agencies; and programmatic environmental impact statements for wind, solar, and geothermal energy. We also obtained funding and staffing data for BLM from the Department of the Interior's Financial and Business Management System. We assessed the reliability of the data we used in our report by reviewing the methods of data collection and data entry into this system, as well as various agency planning documents, and determined that the data were sufficiently reliable to use in this report.

To address our third objective, we reviewed the results of our questionnaire and semistructured interviews regarding both factors that facilitate and hinder renewable energy development. We also interviewed BLM and other agency officials, as well as representatives from seven industry and two environmental groups, to obtain their perspectives on the permitting process, the time frames and factors associated with processing permit applications, and renewable energy development in general.

We conducted this performance audit from January 2012 to January 2013 in accordance with generally accepted government auditing standards. Those standards require that we plan and perform the audit to obtain sufficient, appropriate evidence to provide a reasonable basis for our findings and conclusions based on our audit objectives. We believe that the evidence obtained provides a reasonable basis for our findings and conclusions based on our audit objectives.

APPENDIX II. AUTHORIZED RENEWABLE ENERGY PROJECTS FOR WHICH APPLICATIONS WERE SUBMITTED AFTER THE ENERGY POLICY ACT OF 2005

Of the 445 applications BLM received for utility-scale wind, solar, and geothermal projects since enactment of EPAct 2005 through May 2012, BLM authorized 25 projects—7 wind, 10 solar, and 8 geothermal (see table 2). These projects were authorized with the potential to generate about 5,450 megawatts: 800 megawatts from wind, about 4,200 megawatts from solar, and about 450 megawatts from geothermal energy.

Table 2. Information about Authorized Renewable Energy Projects for Which Applications Were Submitted after EPAct 2005 through May 2012

Energy type	State	Project name	Potential output (megawatts)	Acres	Year authorized	Status
Wind	AZ	Dry Lake Wind Project	63	13,438	2009	In operation
	CA	Tule Wind Energy Project	124	12, 239	2012	Other[a]
	CA	Ocotillo Express Wind Energy Project	315	10,151	2012	In operation
	NV	Spring Valley Wind Project	150	8,600	2010	In operation
	OR	Lime Wind Project	3	107	2010	In operation
	UT	Milford Wind Project, Phase 1	80	110	2008	In operation
	UT	Milford Wind Project, Phase 2	49	34	2010	In operation
Solar	CA	Blythe Solar Power Project	1,000	7,025	2010	Other[b]
	CA	Calico Solar Project	664	4,604	2010	Other[c]

Table 2. (Continued)

Energy type	State	Project name	Potential output (megawatts)	Acres	Year authorized	Status
	CA	Lucerne Valley Solar Project	45	421	2010	Terminated
	CA	Desert Sunlight Solar Farm Project	550	4,144	2011	Under construction
	CA	Genesis Solar Energy Project	250	1,950	2010	Other[d]
	CA	Imperial Valley Solar Project	709	6,459	2010	Terminated
	CA	Ivanpah Solar Electric Generating System	370	3,471	2010	Under construction
	NV	Amargosa Farm Road Solar Project	464	6,320	2011	Terminated
	NV	Crescent Dunes Solar Project	110	2,094	2010	Under construction
	NV	Silver State Solar Energy Project (North)	50	618	2010	In operation
Geothermal	NV	Coyote Canyon Geothermal Project	62	3,960	2011	Other[e]
	NV	Salt Wells Ormat Geothermal Project	40	6,948	2011	Other[e]
	NV	Salt Wells Gradient Geothermal Project	120	15,622	2011	Other[e]
	NV	McGinness Hills Geothermal Project	90	7,680	2011	In operation
	NV	Wild Rose Geothermal Project	35	13,800	2012	Other[f]
	NV	Blue Mountain Geothermal Power Plant	49	5,252	2008	In operation
	NV	Jersey Valley Geothermal Project	30	7,460	2010	In operation
	UT	Cove Fort Geothermal Project	30	50	2012	Under construction

Sources: GAO analysis of questionnaire results and BLM data.

[a] Right-of-way authorized.

[b] Right-of-way holder filed for bankruptcy, and new owner is seeking to change from concentrated solar power technology to photovoltaic solar technology.

[c] Right-of-way was authorized, but applicant is seeking to change technology from sterling dish suncatcher technology to photovoltaic solar technology.

[d] Project was under construction, but work was halted because of cultural resource concerns.

[e] This project has not yet begun construction because the proponent is exploring for additional geothermal resources.

[f] Construction has not begun given the recent authorization at the time of our review.

End Notes

[1] Pub. L No. 109-58, § 211, 119 Stat. 660 (2005). The act specifically omitted hydropower as a part of the 10,000-megawatt goal. A megawatt is 1 million watts, or enough electricity to power about 750 homes.

[2] For this report, we are defining renewable energy to include onshore wind, solar, and geothermal energy. Likewise, we are limiting our discussion of federal lands to those lands managed for multiple uses by the Department of the Interior and the Department of Agriculture. Other lands, such as those managed by the Department of Defense, are excluded from our analysis because they are generally not open to outside development of renewable energy resources.

[3] We interviewed BLM officials from California, Colorado, Nevada, and Wyoming. We selected California, Nevada, and Wyoming primarily because they represent areas with substantial renewable energy development and substantial agency resources devoted to processing renewable energy applications. We selected Colorado primarily to obtain perspectives about the program from a state with less activity and fewer resources.

[4] We interviewed Fish and Wildlife Service officials from the agency's Mountain-Prairie and Pacific Southwest Regions, both selected because they had a significant amount of renewable energy development, and from one state office within each of these regions.We interviewed a National Park Service official from a national park in California because of his experience with renewable energy development near the park.

[5] Other renewable energy sources include bioenergy, such as liquid biofuels and solid biomass fuel; hydropower; ocean energy, including wave, tidal, current, and ocean thermal energy; and waste conversion, including anaerobic digestion, landfill gas, and municipal solid waste.

[6] The National Wildlife Refuge System Improvement Act of 1997 directs the Fish and Wildlife Service to administer a national network of lands and waters for the conservation, management, and, where appropriate, restoration of the fish, wildlife, and plant resources and their habitats for the benefit of present and future generations. The act requires the Fish and Wildlife Service to determine the compatibility of activities with the purposes of a particular refuge and the mission of the refuge system and not allow those activities deemed incompatible. The National Park Service Organic Act of 1916 created the National Park Service to promote and regulate the use of the National Park System with the purpose of preserving unimpaired the natural and cultural resources of park units for the enjoyment of this and future generations.

[7] The Bureau of Indian Affairs is responsible for the administration and management of 55 million surface acres held in trust by the United States for Indian tribes, individuals, and Alaska Natives. Permitting renewable energy projects on tribal trust lands is project dependent and handled by a combination of the tribal surface owner, the bureau, BLM, and other agencies.

[8] Electricity generated from federal lands helps power millions of homes across the United States. For example, the geothermal resources found at the Geysers Geothermal Field—approximately 40 percent of which are considered to come from federal lands—has a net generating capacity sufficient to power 750,000 homes, or a city the size of San Francisco.

[9] The other type of wind turbine is a vertical-axis turbine, which resembles an eggbeater.

[10] In addition, a technology referred to as "enhanced geothermal systems," being developed by the Department of Energy and others, has the potential to allow for the use of geothermal resources not previously obtainable. This technology works by injecting high-pressure fluids into geologic formations where there is hot rock but little natural permeability for hot water to circulate; these fluids fracture the rock and thereby create more permeable formations, enhancing the availability of geothermal resources.

[11] Stakeholders have characterized authorizations for wind and solar energy development as "permits," "rights of way," "right-of-way leases," and "leases." In this report, we use the statutory term "right-of-way" to refer to these authorizations. The Forest Service implements this authority by issuing special use permits.

[12] The Mineral Leasing Act of 1920, as amended, and the Mineral Leasing Act for Acquired Lands of 1947, as amended, provide the legislative authority for federal oil and gas leasing. BLM's oil and gas leasing regulations are located at 43 C.F.R. pt. 3120, and regulations governing oil and gas operations are located at 43 C.F.R. pt. 3160. BLM cannot issue leases for National Forest System lands over the objection of the Forest Service.

[13] BLM may issue a right-of-way on a given land area unless otherwise directed by law, agency order, or a land use plan. 43 C.F.R. § 2802.10.

[14] BLM's website offers proponents a template to be used in developing wind and solar plans of development.

[15] The Federal Land Policy and Management Act authorizes BLM to recover the costs of processing right-of-way permit applications. Such cost-recovery fees are estimated by BLM and can be paid in installments, depending on the actual amount necessary to process an application.

[16] Pub. L. No. 91-190, 83 Stat. 852 (1970), codified as amended at 42 U.S.C. §§ 4321-4347 (2011). Under NEPA, federal agencies must assess the effects of major federal actions—those they propose to carry out or to permit—that significantly affect the environment.

[17] A BLM decision to deny an application must be made in writing, stating the reason(s) for denial. Reasons for denial are described at 43 C.F.R. §2804.26.

[18] Special use authorization regulations appear in 36 C.F.R. pt. 251, subpart B. All uses of Forest Service lands other than those involving roads, grazing and livestock use, the sale and disposal of timber and certain other forest products, and minerals are designated as special uses.

[19] Agency officials told us that limited exploration activities may take place before a lease is issued, although developers generally prefer to obtain a lease before making any substantial investments in exploration activities. The lease is a contract between the federal government and the lessee that specifies certain terms for development and payment of rents and royalties.

[20] Pub. L. No. 91-190 (1970), codified at 42 U.S.C. §§ 4321-4347.

[21] The President's Council on Environmental Quality coordinates federal environmental efforts and works with other White House offices and federal agencies in the development of environmental policies and initiatives.

[22] An environmental assessment is a concise public document that provides sufficient evidence and analysis for determining whether to prepare an environmental impact statement or a

finding of no significant impact and is to include brief discussions of the need for the proposal, alternatives, the environmental impacts of the proposed action and alternatives, and a listing of agencies and persons consulted. 40 C.F.R. § 1508.9 (2011).

[23] An environmental impact statement is a more detailed statement than an environmental assessment, and NEPA implementing regulations specify requirements and procedures—such as providing the public with an opportunity to comment on the draft document—applicable to the environmental impact statement process that are not mandated for environmental assessments. An environmental impact statement must, among other things, (1) describe the environment that will be affected, (2) identify alternatives to the proposed action and identify the agency's preferred alternative, (3) present the environmental impacts of the proposed action and alternatives, and (4) identify any adverse environmental impacts that cannot be avoided should the proposed action be implemented. 42 U.S.C. § 4332(c) (2011), 40 C.F.R. §§ 1501.4, 1508.11 (2011).

[24] A programmatic environmental impact statement does not typically replace the need for site-specific environmental review of each project, but the ability to use broader programmatic analyses can help streamline environmental review by focusing the analysis on the most critical site-specific issues of concern. Known as tiering, a programmatic approach allows an agency to avoid duplication of paperwork by incorporating by reference the general discussions and relevant specific discussions from an environmental impact statement of broader scope into one of lesser scope or vice versa.

[25] Pub. L. No. 93-205 (1973), codified as amended at 16 U.S.C. §§ 1531-1544.

[26] Pub. L. No. 89-665 (1966), codified as amended at 16 U.S.C. §§ 470 to 470x-6.

[27] The act defines as a historic property any prehistoric or historic district, site, building, structure, object, or any properties of traditional religious and cultural importance to an Indian tribe, included or eligible for inclusion in the National Register of Historic Places.

[28] 16 U.S.C. §§ 703-712.

[29] 16 U.S.C.A. §§ 668- 668d.

[30] The one wind development application to the Forest Service since EPAct 2005 was accepted in August 2008 for development in the Huron-Manistee National Forests in Michigan. In 2010, however, before the Forest Service determined whether to approve or deny the permit, the applicant withdrew the application. Since EPAct 2005, additional applications may have been submitted to the Forest Service, but the Forest Service maintains data only for those applications that complete the agency's screening process and are determined sufficiently complete to analyze. This practice differs from that of BLM, which tracks all applications, whether or not they are complete enough to analyze. Separately, the Forest Service has approved one other application for a wind project to be developed on the lands it manages. This project, located in Green Mountain National Forest in Vermont, was approved in January 2012, although construction for the project had not yet begun as of the time of this report. Because the application was submitted before EPAct 2005, this project is not included in our analysis.

[31] In some cases, multiple applications may be submitted and approved for one project. For example, at one solar project in California, the applicant submitted four applications to address the component land areas of a single project. For geothermal energy development, multiple applications are submitted for each project, covering the three phases of geothermal development—exploration, drilling, and construction.

[32] Before applying for a permit for a utility-scale wind project, applicants generally obtain permits for testing and siting. For projects issued a right-of-way, the additional time to obtain these permits ranged from less than one year to about 3.4 years.

[33] BLM has signed a record of decision for two additional projects (one wind energy and one solar energy) that have the potential to generate an additional 1,800 megawatts from projects sited on federal lands. Nevertheless, because these projects had not been issued a right-of-way as of the time of this report, they are not included in our total of authorized projects.

[34] As noted, for purposes of this report we considered wind and solar projects as approved when a NEPA decision (such as a record of decision) is signed, and we considered a project as *authorized* when a right-of-way is subsequently granted.

[35] Several efforts to promote renewable energy development began before EPAct 2005. For example, in May 2001, the President issued an executive order that agencies should take appropriate actions to expedite projects to increase the production, transmission, or conservation of energy. That same month, the President's National Energy Policy Development Group recommended that the Departments of the Interior, Energy, Agriculture, and Defense work collaboratively to increase renewable energy production. Subsequently, in July 2001, the departments created an interagency task force to address how to increase renewable energy production on federal lands.

[36] BLM identified 17 areas, covering approximately 285,000 acres, as solar energy zones.

[37] Land use plans were included for amendment in all states covered by the wind programmatic environmental impact statement except Arizona and California because separate pending land use plan amendments were being conducted to address wind energy development in those states.

[38] BLM Instruction Memorandum No. 2011-061.

[39] The Interior Board of Land Appeals is an appellate review body that exercises the delegated authority of the Secretary to issue final decisions for the Department of the Interior. Its administrative judges decide appeals from bureau decisions including those related to the use and disposition of federal lands and their resources.

[40] BLM Instruction Memorandum No. 2011-059.

[41] BLM Instruction Memorandum No. 2011-061.

[42] Existing BLM regulations encourage applicants wishing to submit a permit application to hold preapplication meetings with BLM but do not require them to do so.

[43] BLM Instruction Memorandum No. 2011-060.

[44] In December 2012, BLM made approximately 303,900 acres of public land unavailable for mining claims under the temporary rule to protect the solar energy zones identified in the solar programmatic environmental impact statement. 77 Fed. Reg. 74690 (December 17, 2012).

[45] Although BLM refers to this proposed competitive process as leasing, the process will continue to use rights-of-way as the basic permitting approach for wind and solar energy development.

[46] In June 2012, the Department of the Interior's Office of Inspector General issued a review of BLM's renewable energy program, reporting that BLM could generate millions of dollars in additional revenues if it used a competitive bidding process. The report primarily addressed areas of renewable energy related to rental revenue collection, bonding, and monitoring and compliance. See Department of the Interior Office of Inspector General, *Bureau of Land Management's Renewable Energy Program: A Critical Point in Renewable Energy Development*, CR-EV-BLM-0004-2010 (Washington, D.C.: June 12, 2012).

[47] Currently, the only such areas are the solar energy zones designated as a result of the solar programmatic environmental impact statement.

[48] Fish and Wildlife Service, *U.S. Fish and Wildlife Service Land-Based Wind Energy Guidelines* (Washington, D.C.: Mar. 23, 2012).

[49] 76 Fed. Reg. 47354. The agency added two new chapters to the *Forest Service Handbook*, addressing wind energy uses and wildlife monitoring and wind energy sites: *Forest Service Handbook* 2709.11 and 2609.13, chapters 70 and 80, respectively, effective August 4, 2011.

[50] These figures represent BLM obligations.

[51] The Federal Land Policy and Management Act of 1976 authorizes the Secretary of the Interior to recover costs, and the Secretary has delegated this authority to BLM. One purpose of this authority is to prevent the federal government from subsidizing private enterprises by paying all their land use application costs. Nevada Power v. Watt, 711 F.2d 913, 925 (10th Cir. 1983).

[52] BLM provided us with the total amount in cost-recovery fees collected in these years but did not specify the amount collected in each individual year.

[53] 42 U.S.C. § 15873. In 2006, we reported that geothermal applicants faced delays when applying to develop power plants on federal lands, concluding that BLM lacked resources to process leases, amend its land use plans, and process applications, which led to delays in many cases. GAO, *Renewable Energy: Increased Geothermal Development Will Depend on Overcoming Many Challenges*, GAO-06-629 (Washington, D.C.: May 24, 2006).

[54] We provided questionnaire respondents with a list of potential factors related to permitting and asked them to identify those that facilitated and hindered the process for each individual project. See appendix I for a full description of our methodology.

[55] The Fish and Wildlife Service has recommended that developers document in writing their actions to avoid, minimize, and compensate for potential adverse impacts. The Fish and Wildlife Service refers to this documentation as Bird and Bat Conservation Strategies.

[56] This guidance was replaced by the Fish and Wildlife Service's March 2012 wind energy guidelines.

[57] For more information about the increased availability and recent price declines of natural gas in the United States, see GAO, *Oil And Gas: Information on Shale Resources, Development, and Environmental and Public Health Risks*, GAO-12-732 (Washington, D.C.: Sept. 5, 2012).

[58] A renewable portfolio standard provides states with a mechanism to increase renewable energy generation using a cost-effective, market-based approach that is administratively efficient. Such a standard requires electric utilities and other retail electric providers to supply a specified minimum amount of customer load with electricity from eligible renewable energy sources. The goal of such a standard is to stimulate market and technology development so that, ultimately, renewable energy will be economically competitive with conventional forms of electric power. Existing portfolio standards vary from state to state. California, for example, has a renewable energy portfolio standard of 33 percent by 2020, and Arizona has established a goal of 15 percent by 2025.

[59] Enacted on January 2, 2013, the American Taxpayer Relief Act of 2012 extended the Energy Production Tax Credit and modified its requirements, allowing projects to be eligible for the credit as long as project construction begins by a deadline of December 31, 2013. This is a difference from past years, where projects were required to be completed and producing electricity by the deadline to be eligible for the credit.

In: Renewable Energy Development ... ISBN: 978-1-62618-764-1
Editor: Carl Polat

Chapter 2

BUREAU OF LAND MANAGEMENT'S RENEWABLE ENERGY PROGRAM: A CRITICAL POINT IN RENEWABLE ENERGY DEVELOPMENT*

U.S. Department of the Interior, Office of Inspector General

Memorandum

To: Michael L. Pool
Acting Director, Bureau of Land Management

From: Mary L. Kendall
Acting Inspector Gene

Subject: Final Evaluation Report — Bureau of Land Management's Renewable Energy
Program: A Critical Point in Renewable Energy Development
Report No. CR-EV-BLM-0004-2010

---

* This is an edited, reformatted and augmented version of U.S. Department of the Interior, Office of Inspector General, Publication Report No. CR-EV-BLM-0004-2010, dated June 2012.

This memorandum transmits the results of our evaluation of the Bureau of Land Management's (BLM) Renewable Energy Program. Our objective was to assess the effectiveness of BLM's development and management of its renewable energy program.

We found that BLM is poised for a massive expansion of wind and solar projects. BLM has taken aggressive action to increase its processing of renewable energy rights-of-way (ROW) grants. BLM's focus on increasing the number of renewable energy projects, however, has exposed some weaknesses in financial accountability and resource protection including obligations to protect the Government's financial interests by collecting rental revenues, managing the bond process, and by appropriate monitoring and enforcing ROW requirements.

We make nine recommendations to improve the long-term management of BLM's Renewable Energy Program to ensure the accurate and timely collection of rents, for managing the bond process, and for monitoring and enforcing ROW requirements. BLM substantially concurred with all nine of our recommendations and agreed to implement them. Based on BLM's response to the draft report (see appendix 3), we consider recommendations 2 and 3 resolved and implemented and the remaining seven recommendations resolved but not implemented (see appendix 4). Accordingly, no further response to the Office of Inspector General (OIG) on this report is necessary.

The legislation creating the OIG requires that we report to Congress semiannually on all audit reports issued, actions taken to implement our recommendations, and recommendations that have not been implemented.

We appreciate the cooperation and assistance of the BLM staff during our review. If you have any questions regarding this report, please contact me at 202-208-5745.

RESULTS IN BRIEF

The Bureau of Land Management (BLM) is poised for a massive expansion of wind and solar projects. At the time of our review, BLM had 29 of 31 authorized wind projects in operation, and 5 of 9 authorized solar projects under construction. Based on the number of wind-testing projects and wind and solar applications, wind projects may quadruple and solar projects may increase tenfold in the near future.

BLM has taken aggressive action to increase its capacity to process renewable energy grants for rights-of-way (ROW). BLM's focus on increasing the number of renewable energy projects, however, has exposed some weaknesses in its management of these projects. Financial accountability and resource protection are integral to the renewable energy program. These include obligations to protect the Government's financial interests by timely and accurately collecting rental revenues and managing the bonding process, and to protect the country's natural resources by monitoring and enforcing grant requirements. Wind and solar projects can cover thousands of acres and therefore have the potential for biological, cultural, historical, paleontological, archaeological, and visual resource degradation.

Secretary Salazar established the production, development, and delivery of renewable energy as one of the U.S. Department of the Interior's (Department) highest priorities by signing Secretarial Order 3285, "Renewable Energy Development by the Department of the Interior," on March 11, 2009. The order describes the need for strategic planning and a thoughtful, balanced approach to domestic resource development. It also noted the Department's role in coordinating and ensuring environmentally responsible renewable energy production and development.

We reviewed BLM's renewable energy activities to assess the effectiveness of BLM's development and management of its renewable energy program. We found that BLM has opportunities to improve in developing and implementing renewable energy policies. For example, although BLM issued guidance on wind rental payments and wind bonding requirements, it did not establish a process to ensure timely implementation of the guidance. This resulted in a loss of $1.2 million in rental revenues on 22 wind projects from 2009 through 2011 and insufficient bonding by $8.5 million on 14 wind projects (see appendix 2).

In addition, BLM has not developed and implemented Bureau-wide guidance on monitoring and inspecting wind and solar projects or for enforcing compliance with ROW requirements. We also found that BLM could potentially generate millions of dollars in additional revenues if it used a competitive bidding process.

We make nine recommendations to improve the long-term management of BLM's renewable energy program. Our recommendations present opportunities for BLM to accurately collect rental revenues, manage the bonding process, monitor and ensure compliance with ROW requirements, and generate additional revenues by using a competitive bidding process for wind and solar ROW.

INTRODUCTION

We evaluated the Bureau of Land Management's (BLM) renewable energy activities and found that BLM is poised for a massive expansion in renewable energy construction and development of wind and solar projects. At the time of our review, BLM had 29 of 31 authorized wind projects in operation. It also had 137 authorized wind-testing projects that could transition into full development projects, and an additional 47 pending applications. BLM had five of nine authorized solar projects just beginning construction, and another 103 applications pending review.

Objective

Our objective was to assess the effectiveness of BLM's development and management of its renewable energy program.

Background

Secretarial Initiatives

On January 16, 2009, former Secretary Kempthorne signed Secretarial Order No. 3283, "Enhancing Renewable Energy Development on the Public Lands." The Order facilitates the U.S. Department of the Interior's (Department) efforts to achieve the goal that Congress established in section 211 of the Energy Policy Act of 2005. That goal is to approve non-hydropower renewable energy projects on public lands—projects with a total generation capacity of at least 10,000 megawatts (MW) of electricity per hour—by 2015. In the Order, the Department declares its support for issuing permits for environmentally responsible development of wind, solar, biomass, and geothermal resources on public lands. The Assistant Secretary for Land and Minerals Management is responsible for issuing permits for renewable energy projects in accordance with applicable laws and regulations, and for establishing renewable energy coordination offices to support permitting activities. Additional responsibilities under the Order include consistently applying renewable energy policies, recovering costs associated with processing renewable energy applications, and monitoring authorized projects. The Federal Land Policy and Management Act of 1976 (FLPMA) authorizes rental payments and cost recovery.

On March 11, 2009, Secretary Salazar signed Secretarial Order No. 3285, "Renewable Energy Development by the Department of the Interior," establishing the production, development, and delivery of renewable energy as a high priority. The Order describes the need for strategic planning and a thoughtful, balanced approach to domestic resource development. It also outlines the Department's role in coordinating and ensuring environmentally responsible renewable energy production and development.

History and Status of BLM's Wind and Solar Activities

BLM manages 20.6 million acres of public lands in 11 Western States that have commercial wind generation potential. BLM has collected $26 million in wind energy rents ($20.5 million from generating facilities and $5.5 million from testing projects) from 1984 through 2010, starting with California in 1984, Wyoming in 1999, Arizona in 2006, and Utah in 2009. BLM has 31 authorized wind projects. Twenty-nine of these projects are currently operating and have a total installed capacity of 783 MW. The largest operating project covers 7,800 acres and generates 80 MW of electricity. BLM also has 137 authorized wind-testing projects that may transition into energy development projects in the future, and 47 wind applications pending.

BLM also manages more than 20 million acres of public lands that have potential for solar energy development. These lands lie in six Western States. Solar energy projects can provide significant amounts of electricity while emitting virtually no greenhouse gases, but they require large areas of relatively flat land. Some technologies also use substantial amounts of water, a scarce commodity in arid climates where solar resources are the most productive. BLM has nine authorized solar projects with a total generating capacity of 3,682 MW. Five projects are under construction—four in California and one in Nevada. The largest authorized project covers 7,025 acres, which will generate 1,000 MW of electricity. BLM also has 103 solar applications pending.

In response to the Secretary's renewable energy initiatives, BLM established 5 Renewable Energy Coordination Offices that, together, have 71 positions; and 6 Renewable Energy Support Teams with 35 positions. This represents a significant investment in renewable energy resources.

Project Authorization

BLM authorizes grants for rights-of-way (ROW) for wind and solar projects under Title V of FLPMA and Title 43, Part 2800, "Rights-of-Way," in the Code of Federal Regulations. BLM reviews ROW applications using a first-come-firstconsidered process. ROW terms and conditions (called stipulations) contain all the details applicable to a grant. By accepting the grant, the grantee agrees to comply with all of the project's stipulations.

BLM instruction memoranda (IM) establish guidance for the renewable energy program. IM 2009-043, "Wind Energy Development Policy," issued December 19, 2008, guides the Bureau on wind projects. The IM was to expire on September 30, 2010, but BLM extended it to September 30, 2012.

BLM authorizes wind projects using three types of ROW grants:

- Type 1 Site-specific Grant for Testing and Monitoring—A site-specific grant for individual, wind-measuring, meteorological towers and facilities. These are term limited to 3 years with no renewal option.
- Type 2 Project Area Grant for Testing and Monitoring—A wind project grant for a larger site-testing and monitoring area where wind resource information is collected to determine the wind energy resource potential of the area. These grants have a term of 3 years that BLM may renew for an additional 3 years if the grantee submits an application for a Type 3 wind energy development ROW, along with a plan for developing the project.
- Type 3 Development Grant—A development grant for wind energy with associated wind turbines as well as on-site access roads, electrical and distribution facilities, and other support facilities. BLM established the term of wind project grants at 30 years to recognize the overall costs and useful life of wind energy facilities.

IM 2011-003, "Solar Energy Development Policy," issued October 7, 2010, is the current guidance on solar projects. Due to the substantial investments required for solar energy projects and the projected life of these facilities, BLM issues these grants for a term not to exceed 30 years.

Rentals and Fees

Grantees pay rents on wind projects in advance for a full calendar year with the following terms:

- Type 1 Site-specific Grant for Testing and Monitoring—A minimum rental rate of $100 per meteorological tower per year. Grantees can pay in advance for the full 3-year term of the ROW.

- Type 2 Project Area Grant for Testing and Monitoring—The higher of $1,000 per year, or $1 per acre per year. Grantees can pay in advance for the full 3-year term of the ROW.
- Type 3 Development Grant—Based on the anticipated MW capacity of the project. The current rate is $4,155 per MW.

Rents for solar projects include two components: base rent charged on a per acre basis starting when BLM issues the ROW, and an MW capacity fee that is phased in over 5 years once the project starts generating power. BLM bases these rents and fees on the approved MW capacity and the type of solar project.

Source: OIG.

Figure 1. Wind turbines near Palm Springs, CA. Rents for wind rights-of-way vary based upon capacity and type of project.

In addition to rents, all wind and solar energy ROW applications and authorizations are subject to cost recovery fees for Bureau processing and monitoring. After BLM receives and conducts an initial review of the ROW application, it estimates the cost of processing the application and issuing a ROW grant. Both BLM and the applicant sign an agreement regarding the Bureau's recovery of these processing costs.

After processing the application, BLM estimates its costs for monitoring the construction, operation, maintenance, and termination of the project; and for protecting and ensuring the rehabilitation of the lands covered by the ROW grant. BLM and the applicant then sign another cost recovery agreement. With both processing and monitoring fees, BLM establishes fee schedules based on the estimated number of Federal work hours for both activities with fixed dollar amounts for estimates of 1 to 50 hours. For those projects that exceed 50 hours, BLM and the grantee negotiate reimbursable agreements with varying fee amounts based on the estimated processing and monitoring hours.

Bond Requirements

BLM uses bonds to ensure compliance with ROW stipulations and applicable regulations, and to protect the Government against loss, damage, or injury to human health, the environment, or property. Currently, the minimum bond rate for wind projects is $2,000 per meteorological tower for wind-testing authorizations (Type 1 and 2 grants) and $10,000 per wind turbine for development projects (Type 3).

Instead of establishing minimum bond amounts, the solar IM describes the three required components of the bond to cover: 1) environmental liabilities; 2) decommissioning, removal, and disposal of improvements and facilities; and 3) site reclamation, revegetation, and restoration. It also establishes a Solar Energy Bond Review Team to assist BLM's state and field offices in reviewing reclamation cost estimates and developing bond requirements.

Source: BLM.

Figure 2. This is a solar project in California that uses a central power tower in the middle of a circle of mirrors.

FINDINGS

Rental Revenues Collection

We found that BLM lost rental revenues of $1.2 million on 22 wind projects for calendar years 2009 through 2011 because it did not implement increased rental rates (see appendix 2).

In December 2008, BLM increased the annual rental rate for wind projects from $2,365 per MW (established in October 2002) to $4,155 per MW. Since companies pay rents for the calendar year in advance, any changes in rental rates are effective the next billing date after rate revisions. BLM did not issue the revised rate with sufficient time to allow field offices to amend the ROW agreements for 2009 and issue the bills for 2009 with the new rental rate.

Furthermore, BLM did not establish a process to ensure that all field offices implemented the new rate. We found two field offices in California, the Palm Springs-South Coast Field Office and the Ridgecrest Field Office, never implemented the new rate for wind projects we reviewed. The Palm Springs-South Coast Field Office was still using rental rates established as far back as 1997. We determined that if BLM had issued the revised rental rate earlier and implemented the new rate on all 22 wind projects in California, BLM could have collected $404,000 more in rental revenues each year from 2009 through 2011, totaling $1.2 million.

Field office personnel acknowledged that they knew about the change in the rental rate and that they had not revised the rates on these agreements. Ridgecrest Field Office personnel said that there were some questions about how to apply the new rental rate and that they had asked for guidance from the State Office. They did not have an answer from the State Office at the time of our visit.

We issued a Notice of Potential Findings and Recommendations to the district manager for the California Desert District Office with our findings and recommendations to apply the new rental rate to all District wind projects by January 1, 2012, and develop and implement procedures to ensure that any future rate changes will be identified and applied. We also recommended that BLM consult with appropriate policy and legal staff to determine whether the new rental rate might be retroactively applied in order to collect revenues lost from 2009 through 2011. The California Desert District Office responded that it started taking actions to address our recommendations, including reviewing regulations, and it will take appropriate action pending the outcome of its review.

In examining data from the Palm Springs-South Coast Field Office, we found that the office was using a rental rate of $2,630 per authorized MW, $265 more than the $2,365 rate applicable from 2003 through 2008. This may have been acceptable and intended by BLM because the October 2002 policy that established the $2,365 rate cited this as a "minimum rent." We are raising this issue because rate-setting policies after 2002 no longer call the rate a "minimum rate," but simply the "rental fee," which raises the question of whether or not a refund is appropriate after considering the increased rate starting 2009.

Recommendations

1. Develop and implement procedures to ensure collection of the current rental rate on all existing wind energy agreements, and timely issuance and prompt application of future rate changes to all wind projects.
2. Determine whether BLM can make the rental rate from December 2008 retroactive in order to collect wind revenues lost during calendar years 2009 through 2011.
3. Determine whether a refund is appropriate for the Palm Springs-South Coast Field Office companies that were charged a higher rate than the established minimum rate from 2003 through 2008.

Bonds on Wind Projects

Bond Requirements on Wind Projects

BLM does not have consistent bond requirements for wind projects. Bonds protect the Government's interests in case of loss, damage, or injury to human health, the environment, or property. The Bureau does not consistently manage bond information or surety instruments to ensure compliance with ROW stipulations and regulatory requirements. As a result, we found 14 wind projects that were either not bonded or insufficiently bonded by a total of $8.5 million (see appendix 2).

Bonding is critical on wind and solar projects because of the "boom" environment created in the wind and solar renewable energy area with the Federal and state governments' renewable energy initiatives, the availability of funding with the American Reinvestment and Recovery Act of 2009, Department of Energy loan guarantees, and tax incentives at both the Federal and state level. BLM experienced this "boom" through the dramatic increase in renewable energy applications, particularly in solar ROW applications in the California Desert District, where solar applications increased from only 2 in 2004 to 130 in 2008. According to BLM renewable energy personnel, the solar energy initiative hit so fast there was very little time for planning by either BLM or companies, with some companies rushing to submit applications just to place "holds" on BLM land. There were concerns about companies creating a speculative environment by trying to obtain a ROW and then sell or transfer the ROW for profit. The changing technologies and financial commitment required of wind and solar projects also creates a high-risk business environment.

This volatile business environment is evident in existing wind projects. For the 31 authorized wind ROW, 21 have been reassigned or changed their names. Two of these have gone through bankruptcy and subsequent reassignment. Eight of the 21 companies have gone through 3 or more name changes.

The uncertain business environment for solar projects is clearly demonstrated by the recently announced bankruptcy of Solar Millennium AG, a German company. Its American subsidiary, Solar Trust of America, currently holds two solar ROW and has two additional applications under review. Development of Solar Millennium's 1,000 MW Blythe project is currently on hold because the company announced its intent to change the project's technology. Such uncertainties reinforce the importance of appropriate levels of performance and reclamation bonding.

BLM did not require bonds on all wind projects. Prior to 2006, BLM's wind policy stated: "A bond is discretionary by the authorized officer, but will usually be required for wind energy development right-of-way grants to ensure compliance with the terms and conditions of the authorization and the requirements of the regulations, including reclamation." Despite this intent, however, BLM field offices did not enforce a bond requirement since the policy made bonds discretionary.

In 2006, the wind policy was revised to make bonding mandatory on new wind energy development projects. The policy did not require, however, bonding on existing wind projects. Bonds on testing projects (Types 1 and 2 ROW) also remained discretionary. The policy suggested $2,500 bonds for wind turbines but stated the actual amount of the required bond would be determined during the ROW authorization process based on site-specific and project-specific factors.

BLM's current wind policy that was issued in December 2008, requires minimum bonds of $2,000 per meteorological tower on Types 1 and 2 wind testing ROW and $10,000 per wind turbine for development projects. It also states that the amount of the bond will be determined during the ROW authorization process based on site-specific and project-specific factors. The explicit requirement for bonds on existing projects, however, remains discretionary because the policy states: "Existing wind energy right-of-way authorizations requiring amendments *may* include provisions of this IM" [emphasis added].

We found that the Palm Springs-South Coast Field Office did not require bonds on 4 wind projects, 1 of which supports about 460 turbines. The Field Office also obtained bonds that were less than the minimum requirements for 10 other projects. In total, BLM did not bond, or insufficiently bonded, 14 projects by $8.5 million (out of 16 wind projects' files reviewed), leaving BLM at risk to future liabilities for land reclamation and the potential damage to natural resources.

We also found that the Pahrump Field Office in Las Vegas, NV, did not bond five projects that are testing wind energy using meteorological towers. We issued a Notice of Potential Findings and Recommendations to BLM reporting the lack of bonds on two wind-testing projects at the Pahrump Field Office and recommended bonding these projects and identifying any other projects without bonds. The acting state director for Nevada agreed with our findings and recommendations. She also identified three additional projects without bonds and assured us that BLM would obtain bonds from the grant holders.

Also in response to our Notice of Potential Findings and Recommendations, BLM issued new wind policies in April 2011 in an effort to establish bonding requirements for all wind projects.[1] We observed, however, that this policy does not rescind and replace wind policies issued in 2008, but reiterates and provides guidance on them. As a result, some BLM personnel and the public may still believe bonding on existing wind projects is discretionary, which according to BLM's Energy Policy Team manager, was not intended.

Bond Coverage on Wind Projects

We found that current minimum bonding requirements for wind projects might not cover potential reclamation costs. The current wind policy does not provide useful instructions on how to determine the appropriate bond amount for wind projects. Instead of including specific instructions for conducting site-specific analysis and assessment to determine bond amounts for wind projects, the policy establishes a minimum bond amount of $10,000 per turbine. BLM field personnel did not know how this minimum bond amount was developed, and expressed their belief that this rate should be higher. Existing bonds we found covering projects for wind energy development ranged from $171 per turbine (a $12,500 bond to cover 73 turbines) to $50,000 per turbine.

The Government Accountability Office (GAO) reported similar issues in BLM's bonding of oil and gas leases in their February 2011 report[2], saying BLM has not consistently implemented its policies for managing potential oil and gas well liabilities. GAO reported that BLM faces two challenges in this area: 1) the minimum bond amounts, not updated in more than 50 years, may not be sufficient to encourage all operators to comply with reclamation requirements; and 2) limitations with the data system BLM uses to track oil and gas information on public lands restricts BLM's ability to evaluate potential liability and monitor agency performance. GAO recommended that BLM develop a comprehensive strategy to increase minimum bond amounts and improve its data system to better evaluate potential liabilities and agency performance.

We did find examples of promising practices in determining bond requirements. For one project, a realty specialist at the Palm Springs-South Coast Field Office determined that neither the minimum bond amount of $10,000 per turbine nor the existing bond rate of $13,000 per turbine was enough. She researched salvage and reclamation costs on comparable wind projects and determined a bond amount of $50,000 per turbine was appropriate, considering decommissioning and removal of the turbines, access roads, and project facilities, and reclamation of the site. She requested $200,000 in bonds for four wind turbines. The company complied.

A realty specialist at the Ridgecrest Field Office also determined that the $10,000 bond per turbine was not enough. On a project that had eight wind turbines, he developed an estimate of $169,800 ($21,225 per turbine) for the bond by taking into consideration compliance monitoring, decommissioning, removal, and site reclamation for the turbines, foundations, and access roads.

A promising practice on solar projects is the use of bond review teams. These teams assist state and field offices in developing site-specific bond requirements for solar energy projects. BLM's solar policy contains very detailed instructions on requirements for solar bonds and the necessary components for establishing the bond amounts. This process was used on a 1,000 MW solar project in California covering 7,025 acres. The bond review team in the State Office reviewed the company's engineering estimates for decommissioning and reclamation of the project and determined the $1.08 million performance and reclamation bond amount for the initial construction phase of the project. As the project progresses into the power generation phase, a new bond amount will be determined.

Bond Management

We found that BLM is not effectively managing the bond process on wind projects. BLM has an incomplete and inaccurate inventory of bonds for wind projects. We found examples of inaccurate bond information in BLM's land and minerals database Legacy Rehost System, called LR2000, at the field offices we visited. The inaccuracies included incorrect bond amounts and incorrect numbers of meteorological towers and wind turbines.

Field personnel told us that the bonding information is only as good as the data in LR2000, and that sometimes they do not input this data. In an effort to improve managing bonds, BLM added requirements in the April 2011 wind policy for all field offices to review and update bond data on wind projects in LR2000's "Bond and Surety System," and to certify this information to BLM Headquarters annually. BLM also still holds bonds for unidentified, transferred, or closed projects. In the Palm Springs-South Coast Field Office, we found six certificates of deposit (CDs) from 1986 that totaled $62,500. Field office personnel did not know which wind projects the CDs covered and speculated that the projects might have been transferred or closed. They told us they knew about the CDs, but had not done the research necessary to determine their proper disposition. We issued a Notice of Potential Findings and Recommendations to BLM reporting these as unidentified bonds and recommended corrective action. The acting manager for the California Desert District responded that BLM agreed with our findings and recommendations.

Recommendations

4. Issue an updated wind IM that clearly requires bonds on all projects.
5. Reassess the minimum bond amounts for wind projects as well as methods for determining the bond amount, including expanding the use of a bond review team.
6. Track and manage bond information on all renewable energy projects, including the amount of the bond, when BLM requested and received the bond, contact information for the bonded party, the type of bond, and when the bond requires updating.

7. Develop and implement procedures to ensure that when a project is transferred, BLM returns the first bond to the company that obtained it and requests a new bond from the newly assigned company.

Monitoring and Inspection

BLM does not have consistent, Bureau-wide guidance for monitoring and inspecting wind and solar ROW. BLM did not monitor or enforce the changing rental and bond requirements for wind projects. This resulted in lost rental revenues of $1.2 million, insufficient bonding of $8.5 million on 14 wind projects, and incomplete and inaccurate bond data. Bureau-wide guidance will also enable industry to know what to expect when planning and developing renewable energy projects.

Monitoring

Wind and solar projects cover large areas—thousands of acres in some cases. These sites may have biological, cultural, historical, paleontological, archaeological, and visual considerations. There may also be environmental concerns related to wind and solar projects regarding the use of chemicals such as pesticides, herbicides, and hazardous materials. As an example of the potential scale of renewable energy projects, BLM approved the largest solar project in California—the Blythe Solar Power Project—a 1,000 MW project covering 7,025 acres. It will include solar power units and ancillary facilities including buildings, access and maintenance roads, bioremediation areas, wastewater treatment facilities, perimeter fencing, gas pipelines, water wells, fiber optic lines, and power lines. Wind-testing projects can also cover large areas. For example, BLM issued a testing and monitoring ROW for a wind project in Nevada that encompassed 34,456 acres.

Monitoring ROWs provides assurance that the grantee is complying with the grant requirements on both the project and program levels. Individual projects need on-the-ground monitoring for changing circumstances such as amendments to the ROW agreement, changing bond levels, changing rental rates, and evolving project life-cycle phases. Life-cycle phases include ROW issuance, project development, construction, operations (power generation), maintenance, and decommissioning (shutdown). ROW agreements can have extensive project stipulations. For example, one authorized solar ROW agreement contains 51 stipulations covering 14 pages of text. Each stipulation requires monitoring and possible enforcement.

Inspection

Inspection ensures compliance with the terms and conditions of individual ROW grants at the project level. ROW agreements do not specifically cite BLM's inspection, monitoring, or enforcement activities, but do include provisions that the holder will pay monitoring fees for BLM's costs, from construction through termination of the ROW.

Our review of wind project files revealed very little inspection documentation for wind projects and no monitoring plans. One BLM field manager told us that wind projects "run themselves" and do not have many issues requiring monitoring.

During our fieldwork, however, we learned that continued monitoring is important to ensure the project site is properly maintained and safe. Monitoring activities can identify incidents such as collapsed wind turbines, bird deaths as a result of hitting wind turbines, damage from fallen meteorological towers, and oil leaks from hydraulic systems.

We found one incident of an unreported, collapsed wind turbine. The collapsed wind turbine was on the only operational BLM wind project in Wyoming in November 2010. The turbine was 1 of 33 turbines on a 21 MW wind project in operation since 1999. The operator did not report the collapsed turbine to BLM. BLM's headquarters personnel found out about the collapsed turbine from pictures posted on FacebookTM and notified the field office in January 2011. According to BLM, the ROW has changed ownership several times. Personnel from the current company said they were not aware they had to report the incident to BLM even though reporting requirements are clearly spelled out in the ROW.

BLM asked the company to investigate the cause of the collapsed turbine. The company submitted a report in July 2011 summarizing the results of the investigations and analyses conducted by the manufacturer and an independent wind energy consultant. The report concluded that the cause of the turbine collapse was a hydraulic system failure during a power outage, which, combined with high winds, caused the turbine blades to spin faster than their designed speed, thereby overstressing the turbine. The report identified the root cause of the hydraulic system failure as a faulty O-ring that caused significant oil leakage.

BLM's greatest concern was the risk to Native American archaeological resources and the environment. The collapsed wind turbine occurred near 25 archaeological sites (hearths and cairns). Although the collapsed turbine did not damage any of the sites, activities removing the turbine may come within 20 feet of ancient cairns. The collapsed turbine also leaked hydraulic fluids.

The company wanted to delay the cleanup of the collapsed turbine to do more analysis after the 2011 winter, but BLM required the removal by October 2011 because of its concerns related to the archaeological resources and leaking fluids.

Source: BLM.

Figure 3. Collapsed 600 kilowatt turbine on BLM wind project in Wyoming that went unreported for several weeks.

The company's report required maintenance and inspections related to hydraulic components to be performed with extreme care to prevent foreign objects from entering the hydraulic system. The report also recommended that the company manually check key components frequently, and that remedial action be taken if components have failed.

If BLM had conducted periodic site inspections of this project, the oil leaks may have been identified sooner and the hydraulic system failure may have been prevented. We were informed that this project was incorrectly coded in LR2000 in 1999 and, therefore, was never identified as requiring a compliance check.

Monitoring Fees

BLM has no Bureau-wide policy for monitoring and enforcement, leaving field offices to develop their own practices. We found considerable inconsistency in monitoring fees at the field offices we visited. At one field office, we reviewed 17 wind project files and found that BLM did not collect monitoring fees for 9 of the projects. For the other eight wind projects, monitoring fees ranged from $5,000 annually to one payment of $5,373 for the 30-year term of the ROW. In most cases, field office personnel did not document how they determined the monitoring fees. In addition, personnel informed us that the monitoring fees reported in LR2000 were unreliable due to a failure to code them correctly.

At one field office, personnel collected monitoring fees for two wind-testing projects, but the amounts were disparate. For one testing project covering 14,222 acres, the office collected only $354 in monitoring fees based on BLM's fee schedule. For the other testing project covering 34,456 acres, the office collected a monitoring fee of $18,853. The field office staff did not document how this fee was determined.

We also found inconsistent monitoring fees for solar projects. One field office used a flat fee of $50,000 each for two solar projects in Las Vegas (a 50 MW project covering 619 acres and a 484 MW project covering 4,350 acres). Another field office, however, developed detailed estimates of $606,000 for a 1,000 MW project covering 7,025 acres and $579,500 for a 250 MW project covering 1,950 acres in Palm Springs.

Third-party Contracting

We also found BLM is using third-party contracting for monitoring and compliance without any Bureau guidance or management controls over the process.

BLM is using third-party contractors for six solar projects authorized in California and one wind project authorized in Utah. The ROW grantee contracts for and pays the monitoring and compliance contractor. During this time of Government austerity, use of third-party contractors may be a viable solution to budget reductions and staffing shortages. BLM needs to maintain an oversight role, however, since inspecting and monitoring wind and solar projects is primarily BLM's responsibility, and building management controls into the process will decrease inconsistency and reduce the risk of noncompliance.

One potential option may be a self-inspection and self-certification process with consequences for noncompliance. For example, on oil and gas leases, BLM requires industry to self-inspect and the Department has authority to penalize false reporting.

Recommendation

8. Develop and implement Bureau-wide guidance for monitoring and enforcement on solar and wind projects, including but not limited to—

   - a consistent process for developing fees for monitoring projects;
   - procedures to plan, conduct, and document monitoring and enforcement of regulations and stipulations;
   - controls over the use of third-party contractors for monitoring and compliance; and
   - self-reporting and self-certification with consequences for non-compliance.

Competitive Bidding for Wind and Solar ROW

In one case in June 2004, BLM used a competitive bidding process for wind energy development instead of the current first-come-first-considered ROW application process. BLM's Palm Springs-South Coast Field Office accepted bids from five companies for a preferential right to apply for a ROW to construct, operate, and maintain wind energy testing and generating facilities on 285 acres. BLM selected the winning bid of $226,500, or $795 per acre. Ordinarily, BLM would have received nothing up front and then the standard ROW fees after issuing the grant.

We estimated that BLM could potentially generate almost $2.5 million in additional revenues if the $795 per acre was applied to the 3,067 acres of BLM land under wind ROW in this area. Of the 19 wind energy development ROW in this area, 15 are due to expire by the end of 2015. Nationally, BLM has more than 30,000 acres currently under wind ROW and more than 31,000 acres under solar ROW. In addition, BLM has identified almost 21 million acres of public lands with wind energy development potential, and more than 20 million acres that have the potential for solar energy development.

BLM informed us that it has initiated formal rulemaking for a competitive bidding process for wind and solar development. BLM drafted an Advance Notice of Proposed Rulemaking that is currently going through its review process. With the demonstrated success of the Palm Springs-South Coast Field Office, the imminent expiration of several ROW in this area, and the current and future wind and solar energy development potential on BLM land, the timing is right for BLM to take the initiative to obtain fair market value for public lands.

Recommendation
9. Develop and implement Bureau-wide guidance for using competitive bidding on wind and solar ROW.

Conclusion and Recommendations Summary

Conclusion

With the projected increase in wind and solar projects on Government lands, the urgency of protecting the people's interests through monitoring and enforcing ROW requirements will increase substantially. By implementing our recommendations, BLM will appreciably improve its immediate and long-term project management. This will involve few additional resources, but will create greater integrity and accountability within the renewable energy program.

BLM responded to the findings and recommendations in our draft report on April 9, 2012 (see appendix 3). BLM substantially concurred with all nine recommendations and completed two of the recommendations (see appendix 4). BLM plans to complete corrective actions for the remaining seven recommendations by April 2013. The responsible official for Recommendations 1, and 4 through 9, is the Assistant Director, Minerals and Realty Management. The responsible official for Recommendations 2 and 3 is the State Director, California State Office.

Recommendations Summary

1. Develop and implement procedures to ensure collection of the current rental rate on all existing wind energy agreements, and timely issuance and prompt application of future rate changes to all wind projects.

   BLM Response: BLM concurred with this recommendation. BLM has procedures to ensure it is collecting current rental rates on existing

ROW grants for the testing and development of wind sites. BLM issued wind energy policy IM 2009-043 on December 22, 2008, which addresses processing, rental, and bonding wind energy projects. It provides guidance to BLM's state and field offices. Many of the 2009 rental bills had already been sent to existing grant holders pursuant to the previous rental policy, as established by BLM policy IM 2003-020 in October 2002. The updated rental rate, as identified in wind energy policy IM 2009-043, was not issued with the intent of amending existing grant-holder rental bills. After BLM sent the initial bills to the grant holders, BLM provided a rental notice and bill wherein grant holders were required to pay the difference in rents to BLM for the years 2009 through 2011.

BLM is also currently in rulemaking for a competitive wind and solar energy leasing program that will also include requirements for paying current rental fees for all wind and solar energy authorizations.

BLM will establish an implementation plan to ensure BLM offices follow the existing procedures and policy guidance. The plan will include regular conference calls, video broadcasts, and an internal review of records to ensure BLM receives current rental rates on all existing wind energy authorizations.

Office of Inspector General (OIG) Analysis of BLM Response: We consider this recommendation resolved, but not implemented. The recommendation will be referred to the Assistant Secretary for Policy, Management and Budget for tracking its implementation.

2. Determine whether BLM can make the rental rate from December 2008 retroactive in order to collect wind revenues lost during calendar years 2009 through 2011.

   BLM Response: BLM concurred with this recommendation. BLM California State Office, in consultation with the Pacific Southwest Regional Solicitor's Office, determined that a retroactive rental for the calendar years 2009 through 2011 was appropriate. The California State Office sent decision letters in July 2011 to the 21 holders of existing grants regarding the adjusted rental for those years. Sixteen grant holders have paid in full.

   OIG Analysis of BLM Response: We consider this recommendation resolved and implemented. No further action is required.

3. Determine whether a refund is appropriate for the Palm Springs-South Coast Field Office companies that were charged a higher rate than the established minimum rate from 2003 through 2008.

BLM Response: BLM concurred with this recommendation. BLM determined, in consultation with the Pacific Southwest Regional Solicitor's Office, that a refund of rentals collected on existing authorizations charged at a rate higher than the established "minimum"rate from 2003 through 2008 is not appropriate. Those rentals were charged based on BLM policy IM 2003-020 which stated that the rental was a "minimum rent," allowing for a higher rental payment if determined appropriate by BLM.

OIG Analysis of BLM Response: We consider this recommendation resolved and implemented. No further response is required.

4. Issue an updated wind IM that clearly requires bonds on all projects.

   BLM Response: BLM partially concurred with this recommendation. The OIG draft report noted that IM 2011-096 did not rescind and replace BLM's 2008 wind-energy policy guidance. IM 2011-096, however, requires bonds for all wind projects and requires that bonds be entered into LR2000. Under this IM, each BLM state director must submit an annual certification of bond status to the Washington Office.

   The 2011 annual certifications have been submitted to the Washington Office, ensuring that bonds are received on all site testing and development authorizations and entered into LR2000. Where an authorization is missing a bond, a letter of noncompliance has been issued to the holder, which includes a timeframe to provide the bond to BLM. BLM's current rulemaking effort for the competitive wind and solar energy leasing program will also include requirements for bonds on all wind site testing, and wind and solar development authorizations.

   BLM will establish a plan to ensure that existing procedures and policy guidance to the BLM offices are followed, that there is a complete understanding of the policy, that bond information is promptly and accurately entered into LR2000, and that the annual certification required by IM 2011-096 is received from each BLM state director. The plan will include regular conference calls, video broadcasts, and an internal review of records to ensure bonds are received on all wind energy authorizations.

   OIG Analysis of BLM Response: We consider this recommendation resolved, but not implemented. The recommendation will be referred to the Assistant Secretary for Policy, Management and Budget for tracking its implementation.

5. Reassess the minimum bond amounts for wind projects as well as methods for determining the bond amount, including expanding the use of a bond review team.

   BLM Response: BLM concurred with this recommendation. BLM will review the minimum bond requirements for wind projects based upon bonds for recently authorized projects and the methods for determining bond amount, including the use of a bond review team. BLM will issue policy guidance, if appropriate.

   OIG Analysis of BLM Response: We consider this recommendation resolved, but not implemented. The recommendation will be referred to the Assistant Secretary for Policy, Management and Budget for tracking its implementation.

6. Track and manage bond information on all renewable energy projects, including the amount of the bond, when BLM requested and received the bond, contact information for the bonded party, the type of bond, and when the bond requires updating.

   BLM Response: BLM concurred with this recommendation. IM 2011- 096 requires entering bonds for all wind projects into LR2000. This includes entering the information suggested in Recommendation 6. Under this IM, each BLM state director must submit an annual certification of bond status to the Washington Office. The BLM state directors also are required to ensure that bonds are received on all site testing and development authorizations and entered into LR2000. Where an authorization is missing a bond, BLM issued a letter of noncompliance to the holder, which includes a timeframe to provide the bond to BLM.

   Bond information for solar energy development projects is required as a standard practice. BLM's Washington Office has reviewed, prior to approval, every solar energy development project authorized. A component of the review is an evaluation of the bond amount for the development facility. BLM policy IM 2011-003 requires the submission of the bond to BLM prior issuing a Notice to Proceed for project development and is subject to Solicitor review prior to acceptance by BLM.

   BLM will review data entered in LR2000 for both wind and solar authorizations and determine if additional fields are necessary to better track and manage bond information. Additionally, BLM will establish an implementation plan for IM 2011-096 as it relates to wind energy authorizations and ensure bond information is collected for solar energy authorizations. The implementation plan will include conference calls, video broadcasts, and Washington Office internal records reviews.

OIG Analysis of BLM Response: We consider this recommendation resolved, but not implemented. The recommendation will be referred to the Assistant Secretary for Policy, Management and Budget for tracking its implementation.

7. Develop and implement procedures to ensure that when a project is transferred, BLM returns the first bond to the company that obtained it and requests a new bond from the newly assigned company.

   BLM Response: BLM concurred with this recommendation. BLM will develop and implement policy guidance to ensure that it receives new bonds as part of the approval process for an assignment, and that it returns old bonds.

   OIG Analysis of BLM Response: We consider this recommendation resolved, but not implemented. The recommendation will be referred to the Assistant Secretary for Policy, Management and Budget for tracking its implementation.

8. Develop and implement Bureau-wide guidance for monitoring and enforcement on solar and wind projects, including but not limited to—

- a consistent process for developing fees for monitoring projects;
- procedures to plan, conduct, and document monitoring and enforcement of regulations and stipulations;
- controls over the use of third-party contractors for monitoring and compliance; and
- self-reporting and self-certification with consequences for non-compliance.

BLM Response: BLM concurred with this recommendation. BLM noted that the OIG's draft report acknowledges the value of using third-party contractors as a solution to budget reductions and staffing issues. BLM will develop and implement policy guidance for wind and solar authorizations to address monitoring, enforcement, and compliance requirements.

OIG Analysis of BLM Response: We consider this recommendation resolved, but not implemented. The recommendation will be referred to the Assistant Secretary for Policy, Management and Budget for tracking its implementation.

9. Develop and implement Bureau-wide guidance for using competitive bidding on wind and solar ROW.

BLM Response: BLM concurred with this recommendation. BLM has initiated formal rulemaking for a competitive wind and solar leasing program. The Advance Notice of Proposed Rulemaking (ANPR) was published in the Federal Register on December 29, 2011. Public comments on the ANPR are currently under review. The final rule will address competitive wind and solar energy leasing, rental fees, required bonding on wind and solar authorizations, and processing and monitoring fees.

OIG Analysis of BLM Response: We consider this recommendation resolved, but not implemented. The recommendation will be referred to the Assistant Secretary for Policy, Management and Budget for tracking its implementation.

APPENDIX 1. SCOPE AND METHODOLOGY

Scope

Our scope included BLM's wind and solar activities, including issued ROW, policies, and guidance.

Methodology

We did this review from March 2010 through December 2011. We visited six BLM offices where we interviewed renewable energy personnel; reviewed selected wind, solar, and transmission ROW files; and observed field operations. The sites visited were Sacramento, Moreno Valley, Palm Springs, Barstow, and Ridgecrest, CA; and Las Vegas, NV. We also interviewed, in person or by telephone, BLM renewable energy personnel in Washington, DC; Lakewood, CO; Reno, NV; Cheyenne and Rawlins, WY; Prineville, OR; and Safford, AZ. We also interviewed a National Park Service employee regarding coordinating activities with BLM related to BLM's wind and solar activities.

We conducted this evaluation in accordance with "Quality Standards for Inspections" issued by the Council of Inspectors General on Integrity and Efficiency. We believe that the work we performed provides a reasonable basis for our conclusions and recommendations.

APPENDIX 2. POTENTIAL MONETARY IMPACT

Issue	Monetary Impact
Lost Rental Revenue	$1.2 million
Insufficient Bonding on Wind Projects	$8.5 million

APPENDIX 3. BUREAU'S RESPONSE

The Bureau of Land Management's response to the draft report follows.

United States Department of the Interior
BUREAU OF LAND MANAGEMENT
Washington, D.C. 20240
http://www.blm.gov

APR 09 2012

To: Assistant Inspector General for Audits, Inspections, and Evaluations
Through: Marcilynn A. Burke
Acting Assistant Secretary - Land Min\frals Management
From: Robert V. Abbey
Director
Subject: Office of the Inspector General Evaluation Draft Report, Bureau of Land Management's Renewable Energy Program: A Critical Point in Renewable Energy Development (CR-EV-BLM-0004-2010)

Thank you for the opportunity to review and comment on the Office of the Inspector General (OIG) draft Evaluation Report, "Bureau of Land Management's Renewable Energy Program: A Critical Point in Renewable Energy Development" (CR-EV-BLM-0004-2010). The Bureau of Land Management (BLM) concurs with eight of the nine recommendations and partially concurs with the recommendation for updating policy to clearly require bonds for the BLM's renewable energy program. The BLM acknowledges a need to reinforce existing policy and provide additional guidance on resource protection, collection of rental revenues, and wind bonding requirements for renewable energy projects. As explained further in the attachment, the BLM has initiated a rulemaking effort for the competitive wind and solar energy leasing program, which will also address several recommendations made by the OIG in its draft report. Additionally, I am pleased to report that we have completed actions to implement two of the recommendations.

As noted in the draft report, the "BLM is poised for a massive expansion of wind and solar projects." As of April 5, 2012, the BLM has 34 authorized wind energy development projects; 140 authorized wind site testing projects; and an additional 166 pending wind energy development and site testing applications. For solar energy development, the BLM has 10 authorized solar energy projects (one project in Nevada is nearly complete in its construction), in addition to another 76 pending solar energy applications.

The attachment provides a summary of the actions taken or planned by the BLM to comply with the OIG's recommendations, as well as the contact information for the responsible official and the target dates of implementation.

If you have any questions about this response, please contact Ray Brady, National Renewable Energy Coordination Office Manager, at 202-912-7312, or LaVanna Stevenson, BLM Audit Liaison Officer, at 202-912-7077.

Attachment

Response to the Recommendations included in the Office of Inspector General Report, Bureau of Land Management's Renewable Energy Program: A Critical Point in Renewable Energy Development (CR-EV-BLM-0004-2010)

Recommendation 1: Develop and implement procedures to ensure collection of the current rental rate on all existing wind energy agreements and timely issuance and prompt application of future rate changes to all wind projects.

Response: The BLM currently has procedures in place to ensure the collection of current rental rates on existing wind site testing and development right-of-way grants. The BLM issued wind energy policy IM 2009-043 on December 22, 2008, which addressed the processing, rental and bonding of wind energy projects. It was issued in an effort to provide guidance to the BLM State and Field Offices to ensure the continued efficient and timely processing of wind energy applications. Many of the 2009 rental bills had already been sent to existing grant holders pursuant to the previous rental policy, as established by BLM policy IM 2003-020 in October 2002. The updated rental rate, as identified in wind energy policy IM 2009-043, was not issued with the intent of amending existing grant holder rental bills. However, after the initial bills were sent to the grant holders, a rental notice and bill was provided wherein grant holders were required to pay the difference in rental to the BLM for the years 2009 through 2011.

The BLM is also currently engaged in a rulemaking effort for a competitive wind and solar energy leasing program that will also include requirements for the payment of current rental fees, as established by the BLM, for all wind and solar energy authorizations.

The BLM will establish an implementation plan to ensure the existing procedures and policy guidance to the BLM States is followed. The implementation plan will include a suite of actions such as regular conference calls, video broadcasts, and a targeted Washington Office internal review of records to ensure current rental rates are received on all existing wind energy authorizations.

Target Date: December 31, 2012

Responsible Official: Michael D. Nedd, Assistant Director, Minerals and Realty Management

Recommendation 2: Determine whether BLM can make the rental rate from December 2008 retroactive in order to collect wind revenues lost during calendar years 2009 through 2011.

Response: The BLM California State Office, in consultation with the Pacific Southwest Regional Solicitor's Office, determined that a retroactive rental for the calendar years 2009 through 2011 was appropriate. The BLM California sent decision letters in July 2011 to the 21 holders of existing grants regarding the adjusted rental for those years. Collection of retroactive rental is currently ongoing. Sixteen grant holders have paid in full. Standard BLM collection policies and practices are being used for the remaining five grants.

Target Date: Completed

Responsible Official: Jim Kenna, State Director, California State Office

Recommendation 3: Determine whether a refund is appropriate for the Palm Springs-South Coast Field Office agreements that were charged a higher rate than the established minimum rate from 2003 through 2008.

Response: The BLM determined, in consultation with the Pacific Southwest Regional Solicitor's Office, that a refund of rentals collected on existing authorizations charged at a rate higher than the established "minimum" rate from 2003 through 2008 is not appropriate. Those rentals were charged based on BLM policy IM 2003-020 which stated that the rental was a "minimum rent," allowing for a higher rental payment if determined appropriate by the BLM.

Target Date: Completed

Responsible Official: Jim Kenna, State Director, California State Office

Recommendation 4: Issue an updated wind IM that clearly requires bonds on all projects.

Response: The BLM partially concurs with this recommendation. It was observed in the draft report that IM 2011-096 did not rescind and replace the BLM's 2008 wind energy policy guidance, which may have contributed to confusion about bonding requirements. However, IM 2011-096 requires bonds for all wind projects and requires that bonds be entered into the BLM's Legacy Rehost System 2000's (LR2000) Bonds and Surety System. Under this IM, each BLM State Director must submit to the BLM Washington Office an annual certification of bond status.

The 2011 annual certifications have been submitted to the BLM Washington Office, ensuring that bonds are received on all site testing and development authorizations and entered into LR2000's Bonds and Surety System. Where an authorization is missing a bond, a letter of noncompliance has been issued to the holder, which includes a timeframe in which to provide the bond to BLM. The BLM's current rulemaking effort for the competitive wind and solar energy leasing program will also include requirements for bonds for all wind site testing, and wind and solar development authorizations.

The BLM will establish an implementation plan to ensure that existing procedures and policy guidance to the BLM States are followed, that there is a complete understanding of the policy, that bond information is promptly and accurately entered into the Bonds and Surety System, and that the annual certification required by IM 2011-096 is received from each BLM State Director. The implementation plan will include a suite of actions such as regular conference calls, video broadcasts, and a targeted Washington Office internal review of records to ensure bonds are received on all wind energy authorizations.

Target Date: December 31, 2012

Responsible Official: Michael D. Nedd, Assistant Director, Minerals and Realty Management

Recommendation 5: Reassess the minimum bond amounts for wind projects as well as methods for determining the bond amount, including expanding the use of a bond review team.

Response: The BLM will review the minimum bond requirements for wind projects based upon bonds established on recently authorized projects and the methods for determining bond amount, including the use of a bond review team. The BLM will issue policy guidance, if appropriate, by December 31, 2012.

Target Date: December 31, 2012

Responsible Official: Michael D. Nedd, Assistant Director, Minerals and Realty Management

Recommendation 6: Track and manage bond information on all renewable energy projects, including the amount of the bond, when BLM requested and received the bond, contact information for the bonded party, the type of bond, and when the bond requires updating.

Response: The BLM policy IM 2011-096 requires the entry of bonds for all wind projects into the BLM's LR2000 Bonds and Surety System. Recordation of bonds in the BLM's LR2000 Bonds and Surety System includes the entry of information suggested in Recommendation 6. Under this IM, each BLM State Director must submit to the BLM Washington Office an annual certification of bond status. The BLM State Directors also are required to ensure that bonds are received on all site testing and development authorizations and entered into LR2000's Bonds and Surety System. Where an authorization is missing a bond, a letter of non-compliance has been issued to the holder, which includes a timeframe in which to provide the bond to the BLM.

Bond information for solar energy development projects is required as a standard practice for the BLM. The BLM Washington Office has reviewed, prior to approval, every solar energy development project authorized to date. A component of the review is evaluation of the bond amount for the development facility. The BLM policy IM 2011-003 requires the submission of the bond to the BLM prior to the issuance of a Notice to Proceed for project development and is subject to Solicitor review prior to acceptance by the BLM.

To strengthen its existing policies further, the BLM will review data entered in its LR2000 Bonds and Surety System for both wind and solar authorizations and determine if additional fields are necessary to better track and manage bond information. Additionally, the BLM will establish an implementation plan for IM 2011-096, as it relates to wind energy authorizations, and ensure bond information is collected for solar energy authorizations at the appropriate time. The implementation plan will include a suite of actions such as conference calls, video broadcasts, and targeted Washington Office internal record reviews.

Target Date: December 31, 2012

Responsible Official: Michael D. Nedd, Assistant Director, Minerals and Realty Management

Recommendation 7: Develop and implement procedures to ensure that when a project is transferred, BLM returns the first bond to the company that obtained it and requests a new bond from the newly assigned company.

Response: The BLM will develop and implement policy guidance to ensure that it receives new bonds as part of the approval process for an assignment, and that it returns old bonds.

Target Date: December 31, 2012

Responsible Official: Michael D. Nedd, Assistant Director, Minerals and Realty Management

Recommendation 8: Develop and implement Bureau-wide guidance for monitoring and enforcement on solar and wind projects, including but not limited to:

- a consistent process for developing fees for monitoring projects;
- procedures to plan, conduct, and document monitoring and enforcement of regulations and stipulations;
- controls over the use of third-party contractors for monitoring and compliance; and
- self-reporting and self-certifications with consequences for non-compliance.

Response: The OIG report acknowledges the value of using third-party contractors as a solution to budget reductions and staffing issues in the current budget environment. The BLM is committed to strengthening its oversight role and developing management controls as a means to increase consistency, reduce the risk of noncompliance, and strengthen environmental protections.

The BLM will develop and implement policy guidance for wind and solar authorizations that addresses monitoring, enforcement, and compliance requirements.

Target Date: December 31, 2012

Responsible Official: Michael D. Nedd, Assistant Director, Minerals and Realty Management

Recommendation 9: Develop and implement Bureau-wide guidance for using competitive bidding on wind and solar ROW.

Response: The BLM has initiated formal rulemaking for a competitive wind and solar leasing program. The Advance Notice of Proposed Rulemaking (ANPR) was published in the *Federal Register* on December 29, 2011. The public comments on the ANPR are currently under review. The Final Rule is expected to be completed by April 30, 2013. The Final Rule will address competitive wind and solar energy leasing, rental fees, required bonding on wind and solar authorizations, and processing and monitoring fees. The BLM is pleased that the OIG concurs that the timing is right for the BLM to take the initiative to obtain fair market value for the use of the public lands.

Target Date: April 30, 2013

Responsible Official: Michael D. Nedd, Assistant Director, Minerals and Realty Management

APPENDIX 4: STATUS OF RECOMMENDATIONS

In response to our draft report, BLM substantially concurred with all nine of our recommendations and agreed to implement them. BLM provided corrective action plans and an action official for each recommendation (see appendix 3). Therefore, we consider two recommendations resolved and implemented and seven recommendations resolved but not implemented.

Recommendations	Status	Action Required
2 and 3	Resolved and implemented	No further action required.
1, 4, 5, 6, 7, 8, and 9	Resolved, not implemented	The recommendations will be referred to the Assistant Secretary for Policy, Management and Budget for tracking of implementation.

End Notes

[1] Bureau of Land Management, IM 2011-096, Certification of Bonding—Wind Energy Site Testing and Wind Energy Development Authorizations (April 7, 2011). The IM requires bonds on all wind energy testing and development projects. It also requires BLM state directors to certify that all wind energy authorizations in their state were bonded.

[2] Government Accountability Office, GAO-11-292, Oil and Gas Bonds; BLM Needs a Comprehensive Strategy to Better Manage Potential Oil and Gas Well Liabilities (February 2011).

In: Renewable Energy Development … ISBN: 978-1-62618-764-1
Editor: Carl Polat

Chapter 3

ENERGY PROJECTS ON FEDERAL LANDS: LEASING AND AUTHORIZATION*

Adam Vann

SUMMARY

Recent concerns over energy supply and pricing have led some to look increasingly to federal lands as a potential energy source. This report explains the legal framework for energy leasing and permitting for onshore lands subject to the control of the federal government.

The report first reviews the laws and regulations affecting leasing of federal lands for exploration and production of oil, natural gas, and coal, as well as the permits that lessees must obtain in order to explore for and produce these resources.

This leasing process has evolved over the last century under the framework established by the Mineral Leasing Act of 1920 (MLA). Oil, natural gas, and coal leasing and production on federal land pursuant to this act are currently overseen by the Bureau of Land Management (BLM), an agency within the U.S. Department of the Interior (DOI). Federal lands identified in the area's Resource Management Plan (RMP) as amenable to oil, coal, or natural gas exploration and production may be leased by BLM, so long as such activities in that area are not prohibited by statute or regulation.

* This is an edited, reformatted and augmented version of Congressional Research Service, Publication No. R40806, dated February 1, 2012.

Such lands are usually leased to the highest bidder as determined by competitive auction. Leaseholders are generally required to pay both rental fees and royalties (a percentage of the value of produced oil, natural gas, or coal) to the U.S. government.

The report also addresses existing laws and regulations that govern the use of federal lands for renewable energy projects, including geothermal, wind, and solar energy. BLM oversees permitting for these projects.

Geothermal projects are leased in accordance with the requirements of the Geothermal Steam Act of 1970. That act functions similarly to the MLA; lands that are amenable to geothermal projects are leased to the highest qualified bidder. In contrast, wind and solar projects on federal lands are not authorized by leases, but rather by obtaining a right-of-way from BLM. These rights-of-way are issued pursuant to the requirements of Title V of the Federal Land Policy and Management Act (FLPMA), and holders of these rights-of-way must make monthly rental payments to the U.S. government.

This framework provides BLM and the federal government with flexibility to use federal lands to help satisfy the nation's energy needs, while generating revenue for the federal government and protecting environmentally sensitive areas.

INTRODUCTION

A variety of interrelated statutes and agency regulations govern leasing and permitting for energy exploration and production on federal lands.[1] Generally, these projects can be divided into two categories, each of which is governed by its own set of statutes and regulations. The first category is the exploration for and production of fossil fuels, including oil, natural gas, and coal. Oil, natural gas, and coal exploration and production on federal lands are generally governed by the Mineral Lands Leasing Act of 1920 (MLA) as amended.

The Bureau of Land Management (BLM), an agency that is part of the U.S. Department of the Interior (DOI), administers the MLA. Generally, the lessee is authorized to explore for and ultimately produce oil, natural gas, or coal on federal lands in exchange for lease payments and royalties paid to the U.S. government on the production. BLM authorizes exploration activities and production activities for lessees only upon lessees' request for permission to engage in these activities on the leased land.

The second category pertains to renewable energy projects. Under the Federal Land Policy and Management Act of 1976 (FLPMA), solar and wind projects are often undertaken pursuant to a right-of-way or similar property interest. However, geothermal energy projects are considered mineral projects and thus are leased under a separate set of laws and regulations in a manner similar to oil and natural gas project leasing.[2]

Oil, Natural Gas, and Coal Exploration and Production on Federal Lands

History and Background

At the start of the 20th century, private entities could explore, develop, and purchase federal lands containing fossil fuels with relative ease. Resources on these federal lands were transferred to full private ownership pursuant to the terms of the Mining Law of 1872. This process was known as "patenting." Under the patenting process, full ownership of oil lands "could be obtained for a nominal amount."[3]

However, Congress eventually decided that oil and natural gas resources on onshore federal lands should remain under federal ownership. The enactment of the MLA in 1920 ended the private acquisition of title to onshore federal lands by authorizing the Secretary of the Interior to issue permits for exploration and to lease lands containing oil, coal, natural gas and other defense-related minerals.[4] The MLA thus allowed the federal government to maintain ultimate control over these federal lands while leasing them to energy companies. The first half of this report details the legal framework for this leasing process.

Federal Lands Subject to Coal, Oil Natural Gas Leasing

The MLA authorizes the Secretary of the Interior to lease virtually all onshore lands owned by the United States that contain fossil fuel deposits, with the federal government retaining title to the lands.[5] BLM is the agency tasked with leasing these subsurface mineral rights not just for the land BLM controls but also for lands controlled by other federal agencies, including those managed by the U.S. Forest Service.[6] The MLA excludes numerous categories of lands such as national parks and monuments, as well as lands in incorporated cities, towns, and villages.[7] Areas within the National Wilderness Preservation System also cannot be leased, but valid property rights that were already in existence by 1984 are preserved.[8] In addition, although BLM controls the leasing process for subsurface rights to land in the National Forest System, the Secretary of the Interior cannot issue any lease for National Forest System lands reserved from the public domain if the Secretary of Agriculture objects.[9]

The Secretary of the Interior is also authorized to withdraw federal lands so that some or all potential land uses, including use for fossil fuel exploration and production, are not permitted on those lands.[10] A withdrawal involves "withholding an area of Federal land from settlement, sale, location, or entry, under some or all of the general land laws, for the purpose of limiting activities under those laws in order to maintain other public values in the area or reserving the area for a particular public purpose or program."[11] However, limitations on the Secretary's withdrawal authority exist. For example, Congress can make withdrawals through legislation, and the Secretary may not modify or revoke a congressional withdrawal.[12]

Development of Resource Management Plans

Bureau of Land Management

According to its website, BLM administers the leasing of minerals found beneath 258 million surface acres of federal lands that it manages, 57 million surface acres where the mineral rights are federally owned but the surface is in non-federal (mostly private) ownership,[13] and approximately 385 million acres whose surfaces are managed by other federal agencies).[14] The Secretary of the Interior must develop and revise Resource Management Plans (RMPs) for the lands under its purview that consider the present and potential future uses for lands managed by BLM.[15] These RMPs serve as the initial determinant of which lands may be subject to leasing, as all activities performed on BLM lands must be consistent with the RMPs.[16]

The Secretary generally must apply "multiple use" and "sustained yield" principles when developing RMPs.[17] "Multiple use" principles require

> management of the public lands and their various resource values so that they are utilized in the combination that will best meet the present and future needs of the American people; making the most judicious use

of the land for some or all of these resources or related services over areas large enough to provide sufficient latitude for periodic adjustments in use to conform to changing needs and conditions; the use of some land for less than all of the resources; a combination of balanced and diverse resource uses that takes into account the long-term needs of future generations for renewable and non-renewable resources."[18]

"Sustained yield" requires the maintenance of "high-level annual or regular periodic output of the various renewable resources of the lands."[19] In addition, the Secretary is required to provide opportunities for the public and various levels of government to participate in the development of RMPs.[20] This can include procedures such as holding public hearings, when appropriate.[21] Regulations require the preparation of an Environmental Impact Statement (EIS) or an Environmental Assessment (EA) in accordance with the National Environmental Policy Act (NEPA)[22] when producing an RMP.[23]

U.S. Forest Service

The Forest Service manages approximately 193 million acres under the "multiple use" and "sustained yield" policies.[24] The Forest Service develops land management plans for Forest Service lands by considering the desired conditions, objectives, suitability of areas for various uses, and other criteria.[25] As with the BLM planning process, the laws governing Forest Service land management and implementation require public notification and opportunities for public participation.[26] When analyzing Forest Service lands for potential leasing, the Forest Service classifies lands into three categories:

1. lands that will be "[o]pen to development subject to the terms and conditions of the standard oil and gas lease form";
2. lands that will be "[o]pen to development but subject to constraints that will require the use of lease stipulations"; and
3. lands that will be "[c]losed to leasing, distinguishing between those areas that are being closed through exercise of management direction, and those closed by law, regulation, etc."[27]

The Forest Service must also comply with NEPA when analyzing NFS lands for potential leasing.[28] Once the Forest Service has completed its analysis of which NFS lands will be available for leasing, it notifies BLM of its decisions, and subsequently may authorize BLM to lease the lands in question.[29]

The Leasing Process: Oil and Gas

The MLA authorizes both competitive and noncompetitive leasing processes for oil and gas exploration and production. Generally, federal lands go through the competitive leasing process first, and if no bids are received, the land will be offered for noncompetitive leasing.

The process begins when BLM generates a list of lands available for competitive oil and natural gas leasing. Private entities may respond by submitting nominations for parcels to be auctioned.[30] No unit being auctioned may exceed 2,560 acres, except in Alaska, where the maximum unit acreage is 5,760 acres.[31] In addition, each unit must be "as nearly compact as possible."[32]

The Secretary must provide 45 days notice before offering federal lands for leasing, including a 30-day period for receiving public comments after notice is published in the *Federal Register*.[33]

Competitive bidding must be held on a quarterly basis in each state where federal lands are available for leasing.[34] The Secretary may also authorize additional opportunities for bidding if he considers them to be necessary.[35]

Once the public notice requirements have been satisfied, the lands are offered for competitive leasing through an oral auction.[36] A national minimum acceptable bid of $2 per acre applies to the auction.[37] Any bids for less than the national minimum bid must be rejected.[38] A competitive bid constitutes a legally binding commitment and cannot be withdrawn.[39] The MLA requires the Secretary to accept the highest bid from a responsible qualified bidder whose bid meets or exceeds the national minimum acceptable bid.[40]

The winning bidder at a competitive auction must submit the following payments on the day of sale, unless otherwise specified: (1) the minimum bid of $2 per acre; (2) the first year's rental payment; and (3) a processing fee as set forth in BLM's regulations.[41] Then, the balance of the bonus bid, if applicable, is due within 10 working days.[42] The lease is issued within 60 days of payment of the remainder of the bonus bid.[43] The lease is also conditioned upon a royalty payment of at least 12.5% in amount or value of the production that is removed or sold from the leased land,[44] unless the Secretary suspends, waives, or reduces the royalty.[45]

The Noncompetitive Leasing Process: Oil and Gas

If no bids are received at a competitive bidding auction—or if all bids submitted are for less than the national minimum acceptable bid—the land will be offered for noncompetitive leasing within 30 days.[46] This noncompetitive leasing remains available for two years after the competitive bidding auction.[47]

The first qualified person who applies for a noncompetitive lease and pays the $75 application fee is entitled to receive the lease without having to competitively bid.[48] All noncompetitive offers received during the first business day after the last day of the competitive auction are considered to have been submitted simultaneously; in such cases, a lottery determines the lease winner.[49] Unlike competitive bids, noncompetitive offers may be withdrawn by those who submitted them within 60 days of filing the offer if no lease has yet been signed on the government's behalf.[50] As with competitive leases, a noncompetitive lease is conditioned upon payment of a 12.5% royalty in amount or value of the oil or gas removed or sold from the leased land.[51] Additionally, there are minimum and maximum acreage limitations for noncompetitive leases.[52] If these criteria are met, BLM will issue the lease within 60 days of the Secretary identifying a qualified applicant.[53]

If no application for a noncompetitive lease is submitted during the two years that the land is available for noncompetitive leasing, the process for leasing the land will again be a competitive oral auction.[54]

Lease Terms and Conditions: Oil and Natural Gas

General Statutory Restrictions

In addition to the processes affecting *where* leasing can take place, general restrictions on leasing address *who* can lease, as well as *how much* land they can lease. First, federal lands containing oil and gas deposits may only be leased to U.S. citizens, associations of U.S. citizens, corporations organized under U.S. laws or the laws of any state, and municipalities.[55] In addition, citizens of a country that denies similar privileges to U.S. citizens and corporations may not control any interest in federal leases.[56] Also, no entity is permitted to own or control oil or gas leases (including options for such leases) under the MLA in excess of 246,080 acres in any one state other than Alaska.[57] Other aggregate acreage limitations include limitations pertaining to options[58] and to combined direct and associational/corporate stockholder interests.[59]

Payment Terms: Rental Fees and Royalties

Leases are conditioned upon payment to the government of a royalty of at least 12.5% in amount or value of oil or gas production that is removed or sold from the leased land.[60] The Secretary has the power to reduce the royalty rate on a noncompetitive lease if he deems it equitable to do so or if circumstances could "cause undue hardship or premature termination of production" absent such a reduction.[61] For oil and gas leases, the royalty must be paid in value unless the Department of the Interior specifies that a royalty payment-in-kind is required.[62] Once the royalty has been paid, the Secretary is required to sell any royalty oil or gas "except whenever in his judgment it is desirable to retain the same for the use of the United States."[63]

In addition to royalties, leases are conditioned upon payment of annual rental fees.[64] Generally, the rate for the first five years of a lease is $1.50 per acre per year, with the rate increasing to $2 per acre for each additional year of the lease.[65] However, there is some variation in rental fee amounts for certain categories of lands.[66] For leases issued after December 22, 1987, a minimum royalty in lieu of the rental fee is due once oil or gas has been discovered on the leased land.[67] The amount of this minimum royalty is equal to the annual rental fee that would otherwise have been due.[68] Rental payments are not due on acreage for which royalties or minimum royalties are being paid, "except on nonproducing leases when compensatory royalty has been assessed in which case annual rental as established in the lease shall be due in addition to compensatory royalty."[69]

Money received from royalties and rental fees is initially paid into the U.S. Treasury.[70] The terms of the MLA provide that 50% of the funds received are allocated to the state where the land or mineral deposit is located.[71] Forty percent of the funds are allocated to the Reclamation Fund under the Reclamation Act of 1902 for projects that provide water to arid Western states.[72] The remaining moneys are credited to the "miscellaneous receipts."[73] Because Alaska is not served by the Reclamation Fund, 90% of the funds collected from federal leases in Alaska are allocated to the State of Alaska.[74] The Secretary is authorized to waive, suspend, or reduce rental fees and royalties under certain conditions.[75]

Lease Terms, Extensions, and Cancellations

The primary term for competitive and noncompetitive leases is 10 years.[76] Leases can be extended to allow for the continuation of exploration operations or oil or gas production. The existence of an approved cooperative plan can also affect extensions.

First, a lease will be extended for two years if three criteria are satisfied:

1. actual drilling operations began before the end of the primary lease term;

2. actual drilling operations are being "diligently prosecuted" at the end of the primary lease term; and
3. rental fees were paid in a timely manner.[77]

Second, a lease that meets these criteria will be extended "so long as oil or gas is being *produced* in paying quantities."[78] A lease that has been extended because of production does not terminate when production stops, as long as the lessee starts reworking or drilling operations within 60 days after production ceases and conducts them with reasonable diligence during the non-productive period.[79]

Furthermore, if a lease initially extended because of drilling begins yielding oil or gas in paying quantities during the two-year drilling extension, the lease can be extended again.[80]

Finally, lessees may collectively adopt and operate under a cooperative plan for exploration and production in a particular area if the Secretary considers such a plan to be in the public interest.[81] All leases subject to such a plan will be extended if any of the leases covered by the plan qualify for a drilling or production extension.[82]

Any MLA lease can be canceled or forfeited if the lessee fails to comply with the statute, the lease's provisions, or regulations promulgated pursuant to MLA.[83] In some situations the Secretary has the authority to cancel the lease unilaterally, but some circumstances require a judicial proceeding to cancel the lease.[84]

In addition, MLA provides for *automatic* termination "upon failure of a lessee to pay rental on or before the anniversary date of the lease, for any lease on which there is no well capable of producing oil or gas in paying quantities."[85] However, the Secretary may in some circumstances reinstate automatically terminated leases.[86]

Applications for Permits to Drill: Oil and Gas

Bureau of Land Management

Operators[87] must submit an Application for a Permit to Drill (APD) for each oil or gas well.[88] Without an approved APD, operators cannot begin drilling operations or cause surface disturbances that are preliminary to drilling.[89]

In fact, the APD process must begin at least 30 days prior to the commencement of operations.[90]

A complete APD must include the following:

- a drilling plan;
- a surface use plan of operations, including drillpad locations and plans for reclaiming the surface;
- evidence of bond coverage;
- a BLM form; and
- any other information that may be required.[91]

Once BLM receives an APD, it must make available information for public inspection for at least 30 days before it may act on the APD.[92] Another pre-approval requirement is that BLM must prepare an environmental record of review or an environmental assessment.[93] Based on these documents, BLM decides whether an EIS is required.[94]

Within five working days of the end of the public notice period, BLM must choose one of four options:

1. approve the application as submitted;
2. approve the application with modifications and/or conditions;
3. disapprove the application; or
4. delay final action.[95]

BLM must also approve a surface use plan of operations addressing proposed surface-disturbing activities before a permit to drill on lands BLM manages may be granted.[96]

U.S. Forest Service

An approved surface use plan of operations addressing proposed surface-disturbing activities is also required before a permit to drill on Forest Service lands may be granted and before any surface-disturbing operations may begin.[97] The operator must submit its proposed surface use plan of operations to BLM as part of its APD.[98] When the proposal pertains to NFS lands, BLM forwards the proposed surface use plan of operations to the Forest Service for evaluation.[99]

The level of detail required in a proposed plan varies depending upon the "type, size, and intensity of the proposed operations and the sensitivity of the surface resources that will be affected by the proposed operations."[100] When evaluating a proposed surface use plan of operations, the Forest Service must ensure that the proposal is consistent with the approved forest land and resource management plan for that area of land.[101] During the evaluation process, the Forest Service must also comply with NEPA, as well as appropriate Forest Service regulations and policies.[102] In addition, the Forest Service can require that the operator increase the amount of its bond if it "determines [that] the financial instrument held by [BLM] is not adequate to ensure complete and timely reclamation and restoration" of the NFS lands.[103]

Ultimately, the Forest Service must decide among four options:

1. approve the plan;
2. approve the plan "subject to specified conditions";
3. disapprove the plan; or
4. delay the plan because additional time is needed to reach a decision.[104]

Once it has made its decision regarding the proposed surface use plan of operations, the Forest Service forwards the decision to BLM, which conducts the leasing of the relevant land in accordance with its regulations.[105]

The Leasing Process: Coal

All BLM coal leasing is done competitively except in cases where a party holds a "prospecting permit" issued prior to the Federal Coal Leasing Amendments Act of 1976 or where contiguous lands are added to existing leases.[106] The process for coal leasing on federal lands is similar to the process for oil and gas leasing. It is governed by Section II of the MLA as amended.[107] Under the statute as amended (most significantly in the aforementioned Federal Coal Leasing Amendments Act of 1976) and the regulations adopted pursuant to it, coal leasing on federal lands may not commence until the land to be leased has been included in a comprehensive land use analysis.[108] The coal screening aspect of this analysis consists of several determinations: (1) the identification of coal-rich areas with potential for development; (2) evaluation of the land in question to determine whether there are any factors that would make it unsuitable for development; (3) consideration of potential multiple use conflicts; and (4) in cases where the federal government does not control the surface estate, consultation with the surface estate owner regarding development of coal resources.[109]

There are two processes by which federal lands may be leased for coal production. The first is "regional coal leasing," in which BLM selects tracts for leasing as needed to meet regional requirements as outlined by "regional coal teams" composed of BLM officials and interested state and local parties. The second is leasing on application.

Regional Coal Leasing

The regional coal leasing process begins with the establishment of regional coal leasing levels by the Secretary of the Interior, informed by the input of the relevant state governors and other interested parties, after consideration of a number of social, economic, and environmental factors.[110] The next step is the delineation and ranking of tracts acceptable for further consideration for coal leasing.[111] After the completing of tract ranking and selection, BLM produces a regional coal leasing environmental impact statement (EIS) to satisfy the requirements of NEPA.[112] This EIS must consider both the site-specific potential environmental impacts of each potential tract lease and the intraregional cumulative environmental impacts of the proposed leasing and alternatives.[113] After completion of the EIS, BLM will consult with any federal agencies potentially affected by the potential leasing,[114] as well as potentially affected governors and Indian tribes,[115] before adopting a regional lease sale schedule.[116]

The lease sales themselves begin with a determination of the fair market value (FMV) of the tract or tracts to be offered, along with the maximum economic recovery for the tract or tracts.[117] BLM does not accept any bid for less than the FMV as determined.[118] After notice of the lease sale is properly given, BLM fields bids and announces the highest bid, which is awarded the lease if it otherwise satisfies all applicable lessee requirements, although BLM reserves the right to reject any and all bids for any reason.[119]

Leasing on Application

BLM also conducts coal leasing in response to applications from interested parties. The process begins with the submission of an application by a qualified potential lessee to lease for purposes of coal development any BLM lands that have been included in a comprehensive coal land use plan or analysis.[120] The application must contain certain data intended to assist BLM in conducting the environmental analysis needed to satisfy the requirements of NEPA.[121] After this process is completed, BLM determines the FMV and maximum economic recovery for the proposed lease tract,[122] and consults with the same parties with whom consultation is required for regional leasing as described in the previous section of this report.[123] After these requirements are met, the lease sale is conducted in the same manner applicable to regional coal lease sales, as described in the previous section.[124]

Lease Terms and Conditions: Coal

General Statutory Restrictions

As with oil and natural gas leasing, coal tracts may only be leased to U.S. citizens, associations of U.S. citizens, corporations organized under U.S. laws or the laws of any state, and municipalities. Also, no entity is permitted to own or control coal leases with an aggregate acreage in excess of 75,000 acres in any one state or more than 150,000 acres in the United States.[125] Parties who have held a federal coal lease for 10 or more years that has not produced commercial quantities of coal may not acquire any other mineral leases under the Mineral Leasing Act of 1920, as amended.[126]

Coal Lease Terms

BLM-issued coal leases are for initial terms of 20 years, with automatic extension "for so long thereafter as coal is produced annually in commercial quantities from that lease."[127] In addition to rental payments of not less than $3 per acre,[128] lessees are required to make payment to the government of a royalty of at least 12.5% in amount or value of coal that is recovered from the leased land.[129] The Secretary has the authority to determine other applicable terms and conditions of coal leases. All leases are subject to the condition of diligent development and continued operation.[130] Lessees must also furnish bonds sufficient to ensure compliance with the terms and conditions of the lease.[131]

Renewable Energy Projects on Federal Lands

Background

In recent years, concern over the impact of emissions from fossil fuel-fired power plants has resulted in increased interest in renewable energy sources. Because the federal government controls land in areas with topography that is considered suitable for renewable energy projects, BLM has issued leases or rights-of-way for a number of renewable energy projects. This section discusses the relevant statutes and regulations applicable to such projects.

The generally applicable requirements and restrictions for lands managed by BLM discussed previously in this report also apply to this discussion of renewable energy projects on federal lands. Renewable energy uses can be considered during the formulation of Resource Management Plans (RMPs) for federal lands. Because the energy source does not expire and because the requisite gathering facilities often leave a relatively small footprint, renewable energy can often be an attractive option when applying the "multiple use" and "sustained yield" principles during the RMP process as described *supra* in the section of this report titled "Development of Resource Management Plans."

Geothermal Project Leasing

Background

Geothermal energy is energy that is produced from heat stored under the surface of the earth. Geothermal energy is produced by releasing cold water under the surface into a "hot" area in order to generate steam to fuel the generation of power. BLM has the delegated authority to lease 247 million acres of lands in the Western United States and Alaska (including 104 million acres of lands managed by the Forest Service)[132] with geothermal potential.[133] BLM currently manages 530 geothermal energy leases, with 58 leases producing 1,275 megawatts of electricity, about half of the total amount of U.S. geothermal energy production capacity.[134]

Geothermal energy projects administered by BLM differ from most other renewable energy projects on federal lands in that they are operated pursuant to a lease, rather than a more limited authorization like a right-of-way. The governing law for steam leasing is the Geothermal Steam Act of 1970.[135] The act sets forth detailed provisions governing the issuance and administration of geothermal steam leases. The laws and regulations governing geothermal steam leasing and administration are similar to the principles and processes for oil and natural gas leasing on federal lands described in detail earlier in this report.

The Leasing Process

The Geothermal Steam Act, like the MLA, authorizes both competitive and noncompetitive leasing processes. The competitive process begins when BLM receives nominations for parcels to be auctioned from a qualified company or individual.[136] Nominated areas may not exceed 5,120 acres unless the lease area includes an irregular subdivision.[137] Qualified companies or individuals may request that nominated parcels be offered as a block for competitive leasing.[138]

The Secretary must provide 45 days notice by posting in the applicable BLM office before offering federal lands for geothermal leasing.[139] BLM must hold competitive bidding for geothermal leases at least once every two years.[140] Once the public notice requirements have been satisfied, the federal lands are offered for competitive leasing through an auction at the time, date, and place specified in the public notice.[141]

Bidders must make a minimum acceptable bid of $2 per acre for the first year, $3 per acre for years 2-10, and $5 per acre for every year after year 10 of the lease.[142] A competitive bid constitutes a legally binding commitment and cannot be withdrawn.[143] BLM will accept the highest bid from a qualified bidder.[144] The winning bidder must pay 20% of the bid amount, the total amount of the first year's rental, and a processing fee to BLM by the close of business on the day following the day of the lease sale.[145] The remainder of the bid amount must be submitted to BLM within 15 calendar days of the day of the sale.[146]

Lands offered at a competitive lease sale that receive no bids are made available for noncompetitive leasing for a two-year period beginning the first business day following the date of the lease sale.[147] For the first 30 days following the lease sale, leases are available only for parcels as configured for the lease sale; after 30 days, BLM accepts applications for any available lands covered by the lease sale.[148] Holders of mining claims on federal lands may apply for noncompetitive geothermal leases within the lands on which they hold claims.[149]

Geothermal steam project leases can also be obtained though "direct use" leasing. Direct use of geothermal energy involves the use of geothermal hot water directly for its intended end use, rather than for transfer into electricity that is then transmitted to the end user. Applicants can obtain direct use leases if (1) the lands in question are open for geothermal leasing; (2) BLM determines that the lands are appropriate for direct use operations without sale for purposes other than commercial generation of electricity; (3) BLM has published notice of the proposal to use the land for direct use leasing; (4) during the 90-day period following the publication of such notice, BLM does not receive any nomination to include the land in question in the next competitive lease sale; and (5) BLM determines there is not competitive interest in the resource.[150]

Geothermal leases generally have an initial primary term of 10 years, with a number of extensions that can extend the term at the discretion of BLM.[151] The first five-year extension is granted if the lessee has expended a minimum of $40 per acre in development activities that provide additional geologic or reservoir information.[152] Subsequent extensions may be granted upon showings of continued annual expenditures and progress in the development of geothermal resources.[153]

Lessees are required to make annual rental payments to BLM. For leases acquired in a competitive lease sale and issued on or after August 8, 2005, the annual rent is $2 per acre for the first year and $3 per acre for the second through the 10th year.[154] For leases acquired noncompetitively, the annual rent is $1 per acre for the first 10 years of the lease.[155] For both types of leases, the annual rent is $5 per acre each year after the 10th year.

Lessees must also make royalty payments on electricity produced from geothermal steam on federal lands. For leases issued on or after August 8, 2005, the royalty rate is 1.75% of gross sales for the first 10 years of production, and 3.5% thereafter.[156] If the lessee sells the geothermal resources to a third party who then uses those resources to produce electricity, the royalty rate on the sale is 10%.[157] Direct use lessees do not have to pay royalties unless they sell the geothermal steam resources to a third party; instead, they must pay a direct use fee in accordance with a published fee schedule.[158]

Geothermal leases can be suspended or canceled at the discretion of the Secretary of the Interior. The Secretary may suspend leases in the interest of conservation of resources, and may terminate leases for any violation of the BLM regulations or the lease terms.[159] The Secretary is also authorized to waive, suspend, or reduce the rental or royalty rates for any lease in the interest of conservation.[160] The lessee may also suspend the lease on their own motion in the interest of conservation.[161]

Exploration and Production under Geothermal Leases

As with oil and natural gas leasing, a geothermal lease alone does not entitle the lessee to commence drilling and production. In order to conduct exploration for geothermal resources, the lessee must submit to BLM a Notice of Intent to Conduct Geothermal Resource Exploration Operations.[162] Exploration operations must be conducted in a manner that (1) protects the quality of surface and subsurface waters, air, and other natural resources; (2) protects the quality of cultural, scenic, and recreational resources; (3)

accommodates other land uses; and (4) minimizes noise.[163] Data collected during exploration operations must be shared with BLM.[164]

Lessees must also submit a geothermal drilling program and obtain a drilling permit before commencing drilling operations for geothermal resources on leased land.[165] Similar to exploration operations, drilling operations must be conducted in a matter that (1) protects the quality of surface and subsurface waters, air, and other natural resources; (2) protects the quality of cultural, scenic, and recreational resources; (3) accommodates other land uses; (4) minimizes noise; and (5) prevents property damage and unnecessary or undue degradation of the lands.[166] Lessees must also meet BLM's geothermal well abandonment requirements upon completion of operations.[167]

Authorizations for Wind and Solar Energy Projects

While oil, gas, and geothermal projects are permitted by BLM under leasing processes, those interested in producing wind or solar energy on federal lands do not seek leases. Instead, they seek more limited authorizations for development of their energy projects pursuant to Title V of the Federal Land Policy and Management Act (FLPMA).[168]

Background

Federal lands also have significant capacity for both wind and solar energy production. BLM manages 20.6 million acres of federal lands with wind potential in 11 Western states.[169] BLM has amended 52 different land use plans to allow for development of wind energy projects.[170] The Southwestern United States has some of the highest solar radiation levels in the world, a characteristic that makes it a possible candidate for solar energy development.[171] Because of these characteristics, BLM has been active in consideration and promotion of wind and solar energy projects on the lands that it manages.

As with oil, gas and geothermal leasing, not all federal lands are available for wind and solar renewable energy project rights-of-way. Lands designated as Wilderness Areas and Wilderness Study Areas, National Monuments, National Conservation Areas (with the notable exception of the California Desert Conservation Area), National Wild and Scenic Rivers, and National Historic and Scenic Trails, are categories of land not open to solar and wind energy development.[172] In addition, some special management areas, such as

Areas of Critical Environmental Concern, may not be suitable for development.[173]

Title V of FLPMA authorizes BLM and the Secretary of Agriculture (managing Forest Service lands) to "grant, issue or renew rights-of-way over, upon, under or through" their administered lands for "systems for generation, transmission and distribution of electric energy."[174] Neither Title V of FLPMA nor the accompanying regulations address wind or solar energy projects specifically. However, because wind and solar projects on federal lands are conducted via rightsof-way, many of the provisions of FLPMA and the accompanying regulations are relevant.

The agencies may exercise their discretion to grant rights-of-way on any lands under their jurisdiction, unless (1) prohibited by a statute, regulation or order that specifically excludes rights-of-way; (2) the lands are specifically segregated or withdrawn from the use contemplated by the right-of-way; or (3) the agency has identified the area in land use plans as inappropriate for right-of-way uses.[175]

BLM Rights-of-Way

Parties applying for BLM rights-of-way for development of wind or solar energy projects must satisfy a number of requirements in their application, as set forth by the Secretary of the Interior.[176] These include disclosures regarding information related to the use of the right-of-way, information regarding the identification of business partners and other affiliated entities, and information regarding plans for use of the right-of-way.[177] Applications for rights-of-way on BLM lands must be accompanied by a processing fee as set forth in BLM regulations.[178]

Rights-of-way granted by BLM function differently from leases in several respects. Perhaps the most significant difference between leases and rights-of-way are the substantial rights to use of the land in question retained by the lessor, the United States. BLM retains the right to

1. access the lands covered by the right-of-way, including any facilities constructed on the right-of-way;
2. require common use of the land (including subsurface and air space) and to authorize others to use the right-of-way for compatible uses;
3. retain ownership of the resources of the land;
4. determine whether or not the grant is renewable; and

5. change the terms and conditions of the right-of-way as a result of changes in legislation or regulation or as otherwise necessary to protect public health or safety or the environment.[179]

BLM may also include in a right-of-way grant any terms, conditions, or stipulations that it determines to be in the public interest, including modifications of proposed routes or changes to the location of certain facilities.[180] Rights-of-way may also completely prohibit use of the land until BLM has approved a Plan of Development for the land and issued a Notice to Proceed to the right-of-way holder.[181] The terms of rights-of-way vary, but those with terms longer than 20 years must include periodic reviews at the end of the 20th year and at 10-year intervals thereafter.[182]

Holders of BLM rights-of-way are also required to make rent payments as set forth by BLM. Rent is payable on a monthly basis,[183] and waivers or reductions can be obtained.[184] BLM regulations provide that rent for non-linear rights-of-way for purposes other than communications uses will be determined through a process based on comparable commercial practices, appraisals, competitive bidding, or other reasonable methods.[185]

In an effort to make the development of renewable energy projects on federal lands more efficient and to ensure fair market value, DOI is seeking public comment on establishing a competitive leasing system for wind and solar projects on federal lands. DOI issued an Advanced Notice of Proposed Rulemaking accepting comments through February 27, 2012.[186] Related legislative action—the Public Lands Renewable Energy Development Act of 2011—would place an emphasis on conducting a two-year pilot leasing program for solar and wind energy projects on public lands and for providing a leasing framework for development should the Secretaries of Agriculture and Interior recommend that a leasing program be established.[187]

Forest Service Special Use Authorizations

The Forest Service also does not have regulations that specifically address wind and solar energy rights-of-way, so the agency would look to generally applicable laws and regulations as guidance for the authorization of wind or solar projects. The Forest Service does not currently authorize any such projects on its lands. Applicants seeking to make use of Forest Service lands for such a project would apply for "special use authorization" under the generally applicable regulations for such authorizations.

All uses of National Forest System lands other than road usage, grazing and livestock use, sale and disposal of timber and other forest products, and mineral usage are designated as "special uses."[188] Before conducting a special use on these lands, individuals or entities must submit a proposal to a Forest Service-authorized officer[189] and must obtain "special use authorization" from that authorized officer, unless that requirement is waived under regulations that are likely not applicable to energy projects.[190] Among the permitted authorizations are "permits, leases and easements ... for rights-of-way for ... systems and related facilities for generation, transmission and distribution of electric energy,"[191] which would authorize wind or solar generation facilities.

Applicants for special use authorizations must contact the Forest Service office(s) responsible for management of the affected land as early as possible in advance of the proposed use.[192] Upon receipt, the authorized officer screens the application in two phases to ensure that the proposed use is consistent with the laws, regulations, orders, and policies establishing or governing National Forest System lands,[193] that the proposed use is consistent with the standards and guidelines in the applicable forest land and RMP,[194] that the proposed use will not create an exclusive or perpetual right of use or occupancy,[195] and that other Forest Service initial screening concerns are satisfied.[196] After this second-level screening, a formal application is submitted and processed,[197] and federal, state, and local agencies and the public are afforded the opportunity to comment on the application.[198] After review and a decision to approve, the authorized officer may issue a special use authorization.[199]

Special use authorizations are subject to a number of limitations. All rights not expressly granted are reserved to the United States, including

1. continuing rights of access to all Forest System lands;
2. a continuing right of physical entry to any part of the authorized facilities; and
3. the right to require common use of the land or to authorize the use by others in any way not inconsistent with a holder's existing rights and privileges after consultation.[200]

A number of other mandatory terms and conditions apply to the Forest Service-issued special use authorizations, as set forth in the regulations.[201] Special use authorization holders also must pay an annual rental fee for the authorization, as determined by the authorized officer based on the fair market value of the rights and privileges authorized, as determined by appraisal or other sound business management principles.[202] The Forest Service also may collect processing fees from applicants and holders to recover its costs.[203]

The Forest Service may suspend, revoke, or terminate a special use authorization under any of the grounds articulated in its regulations.[204] The regulations do not provide specific guidance on duration and renewability of special use authorizations, providing simply that "[i]f appropriate, each special use authorization will specify its duration and renewability. The duration shall be no longer than the authorized officer determines is necessary to accomplish the purpose of the authorization and to be reasonable in light of all circumstances concerning the use."[205] Where renewal is provided for, the authorized officer will renew the authorization if the project or facility is still being used for the purpose(s) previously authorized and is being operated and maintained in accordance with all the provisions of the authorization.[206] Where no such renewal provision is in place, the authorization can be renewed at the authorized officer's discretion upon request of the holder.[207]

End Notes

[1] This report provides a discussion of energy projects on onshore federal lands (i.e., any land in which the federal government maintains a property interest). For discussion of offshore energy projects, see CRS Report RL33404, *Offshore Oil and Gas Development: Legal Framework*, by Adam Vann; CRS Report R40175, *Wind Energy: Offshore Permitting*, by Adam Vann; and CRS Report RL34741, *Drilling in the Great Lakes: Background and Issues*, coordinated by Pervaze A. Sheikh.

[2] The licensing and regulation of hydroelectric facilities on federal lands are not covered in this report. They will be covered in a forthcoming CRS product.

[3] *See* Pan Am. Petroleum & Transp. Co. v. United States, 273 U.S. 456, 486 (1927) (internal citation omitted).

[4] Mineral Lands Leasing Act of 1920 41 Stat. 4373 (1920), codified at 30 U.S.C. §181 *et seq.*

[5] 30 U.S.C. §181.

[6] *Id.*

[7] *Id.*

[8] 16 U.S.C. §1133(d)(3) ("Subject to valid rights then existing, effective January 1, 1984, the minerals in lands designated ... as wilderness areas are withdrawn from all forms of appropriation under the mining laws and from disposition under all laws pertaining to mineral leasing and all amendments thereto.").

[9] *Id.* at §226(h). In addition, the U.S. Forest Service has issued separate regulations governing certain aspects of leasing and permitting for oil and gas development on lands within its jurisdiction.

[10] 43 U.S.C. §1714 (2000).

[11] *Id.* at §1702(j).

[12] 43 U.S.C. §§1701(a)(4), 1714(j); 43 C.F.R. §2300.0-3(a).

[13] http://www.blm.gov/wo/st/en/prog/energy

[14] *Id.* In addition to the Forest Service lands discussed in this report, the 385 million acres for which BLM would manage subsurface rights include National Park lands, lands controlled

by the Fish and Wildlife Service or the Department of Defense, and other lands that are not frequently leased for development of subsurface natural resources.

[15] 43 U.S.C. §1712; 43 C.F.R. Parts 1600, 1610.

[16] 43 C.F.R. §1610.5-3.

[17] 43 U.S.C. §§1701(a)(7), 1732(a). An exception applies in circumstances "where a tract of such public land has been dedicated to specific uses according to any other provisions of law [in which case] it shall be managed in accordance with such law." 43 U.S.C. §1732(a).

[18] *Id.* at §1702(c).

[19] *Id.* at §1702(h).

[20] *Id.* at §1712(f).

[21] *Id.*

[22] 42 U.S.C. §§4321 et seq.

[23] *See* 43 C.F.R. §1601.0-6. The regulations stipulate that approval of an RMP constitutes a "major Federal action significantly affecting the quality of the human environment"; thus, the preparation of an Environmental Impact Statement (EIS) is required pursuant to NEPA. *Id.*; *see* 42 U.S.C. §4332(2)(C). NEPA also requires BLM to take a "hard look" at the environmental impacts of significant proposed actions. 42 U.S.C. §4332(2)(C). If it is unclear whether a proposed action will have significant environmental effects, the agency may prepare an Environmental Assessment (EA). 40 C.F.R. §1501.3. Based on the EA, an agency may issue a finding of no significant impact (FONSI), thereby concluding the NEPA process, or it may determine that preparation of an EIS is necessary. *Id.* §1508.13. An EIS must address numerous issues, including the environmental impact of the proposed action, alternatives to the proposed action, and any irreversible commitments of resources which would be involved if the proposed action takes place. 42 U.S.C. §4332(2)(C). For further discussion of NEPA, see CRS Report RS20621, *Overview of National Environmental Policy Act (NEPA) Requirements*, by Kristina Alexander.

[24] Multiple-Use Sustained-Yield Act, P.L. 86-517, codified at 16 U.S.C. §§528-532.

[25] 16 U.S.C. §1604; 36 C.F.R. §§219.5, 219.7

[26] 16 U.S.C. §§1604(d), 1612; 36 C.F.R. §219.9.

[27] 36 C.F.R. §228.102(c).

[28] 16 U.S.C. §1604(g); 36 C.F.R. §228.102(a).

[29] 36 C.F.R. §228.102(d), §228.102(e).

[30] *See* 43 C.F.R. §3120.3-1.

[31] 30 U.S.C. §226(b)(1).

[32] *Id.*

[33] 30 U.S.C. §226(f).

[34] 30 U.S.C. §226(b)(1).

[35] *Id.*

[36] *Id.*

[37] 30 U.S.C. §226(b)(1)(B); 43 C.F.R. §3120.1-2.

[38] 30 U.S.C. §226(b)(1).

[39] 43 C.F.R. §3120.5-3(a).

[40] 30 U.S.C. §226(b)(1).

[41] 43 C.F.R. §3120.5-2.

[42] *Id.*

[43] 30 U.S.C. §226(b)(1).

[44] *Id.*

[45] 30 U.S.C. §209.

[46] 30 U.S.C. §226(b)(1), (c).
[47] *Id.*
[48] 30 U.S.C. §226(c)(1).
[49] 43 C.F.R. §3110.2(a).
[50] *Id.* at §3110.6.
[51] 30 U.S.C. § 226(c)(1). As mentioned above, the Secretary may suspend, waive, or reduce rentals or royalties under certain conditions. *See, e.g.*, 30 U.S.C. §209.
[52] *See* 43 C.F.R. §3110.3.
[53] 30 U.S.C. §226(c)(1).
[54] 30 U.S.C. §226(c)(2)(A).
[55] 30 U.S.C. §181.
[56] *Id.*
[57] *Id.* at §184(d)(1). For Alaska, the limit is 300,000 acres in the northern leasing district and 300,000 acres in the southern leasing district, of which no more than 200,000 acres may be held under options in each of the two leasing districts. 30 U.S.C. § 184(d).
[58] 30 U.S.C. §184(d)(2).
[59] *Id.* at §184(e)(1).
[60] *Id.* at §226(b)(1).
[61] 30 U.S.C. §188(i)(1).
[62] 30 C.F.R. §§202.100, 202.150.
[63] 30 U.S.C. §192.
[64] *Id.* at §226(d).
[65] *Id.*; 43 C.F.R. §3103.2-2(a). The rental payment for the first year of the lease must be included with each competitive bid or noncompetitive lease offer. 43 C.F.R. §3103.2-1.
[66] *See* 43 C.F.R. § 3103.2-2.
[67] *Id.* at §3103.3-2(a).
[68] *Id.*
[69] *Id.* at §3103.2-2(c).
[70] *Id.* at §191(a).
[71] *Id.* It should be noted that for the past several years, appropriations legislation for the Department of the Interior has provided that 2% of amounts payable to the states for the given fiscal year under the MLA should be deducted and deposited to the Treasury. See, e.g., P.L. 112-74, Division E, Title I, under the heading of "Departmental Operations."
[72] *Id.*
[73] *Id.*
[74] *Id.*
[75] *See, e.g.*, 30 U.S.C. §209.
[76] 30 U.S.C. §226(e).
[77] *Id.*; 43 C.F.R. §3107.1. Other regulations regarding the continuation, extension, and renewal of leases are set forth in 43 C.F.R. §§3107.1 *et seq.*
[78] 30 U.S.C. §226(e) (emphasis added).
[79] 30 U.S.C. §226(i); 43 C.F.R. §3107.2-2.
[80] *See* 30 U.S.C. §226(e).
[81] *Id.* at §226(m).
[82] *Id.*; 43 C.F.R. §3107.3-1.
[83] 30 U.S.C. §188(a); 43 C.F.R. §3108.3(a), (b).
[84] *See* 43 C.F.R. §3108.3(a), (b).
[85] 30 U.S.C. §188(b).

[86] *See id.* at §188(c)-(e), (g), (j).
[87] BLM Regulations define an *Operator* as "any person or entity including but not limited to the lessee or operating rights owner, who has stated in writing to the authorized officer that it is responsible under the terms and conditions of the lease for the operations conducted on the leased lands or a portion thereof." 43 C.F.R. §3160.0-5.
[88] *Id.* at §3162.3-1(c); *see* 43 U.S.C. §1732(b).
[89] 43 C.F.R. §3162.3-1(c).
[90] *Id.* at §3162.3-1(d).
[91] *Id.* at §3162.3-1(d), (f).
[92] 30 U.S.C. §226(f); 43 C.F.R. §3162.3-1(g).
[93] 43 C.F.R. §3162.5-1(a).
[94] *Id.*
[95] 43 C.F.R. §3162.3-1(h).
[96] 30 U.S.C. §226(g).
[97] 30 U.S.C. §226(g); 36 C.F.R. §228.106(a).
[98] *Id.*
[99] *Id.*
[100] 30 U.S.C. §228.106(c).
[101] 16 U.S.C. §1604(i); 36 C.F.R. §228.107(a)(2).
[102] 36 C.F.R. §228.107(a).
[103] *Id*. at §228.109(a).
[104] *Id.* at §228.107(b).
[105] *Id.* at §228.107(d).
[106] *Id*. at §203.
[107] 30 U.S.C. §201 *et seq.*
[108] *Id*. at §201(a)(3)(A)(i).
[109] 43 C.F.R. §3420.1-4.
[110] *Id.* at §3420.2.
[111] *Id.* at §§3420.3-3; 3420.3-4.
[112] *Id*. at §3420.3-4(c).
[113] *Id.*
[114] *Id* at §3420.4-2.
[115] *Id*. §§3420.3-3; 3420.4-4.
[116] *Id*. at §3420.5.
[117] *Id*. at §3422.1.
[118] *Id.* at §3422.1(c)
[119] *Id*. at §§3422.3-2; 3422.4
[120] Id. at §§3425.1, 3425.2.
[121] id. at §§3425.1-7; 3425.3.
[122] *Id*. at §3424.4(a)(1).
[123] *Id*. at §3424.4(a)(2).
[124] *Id*. at §3424.4(b).
[125] 30 U.S.C. §184(a).
[126] *Id*. at §201(a)(2).
[127] *Id*. at §207(a).
[128] 43 C.F.R. §3473-3.1(a).
[129] 30 U.S.C. §207(a).
[130] *Id*. at §207(b)(1).

[131] 43 C.F.R. §3474.2.

[132] According to the Geothermal Steam Act, the Secretary of the Interior may issue leases for developments and utilization of geothermal resources in any federal lands administered by BLM, or in national forests administered by the Forest Service, or in lands conveyed by the United States if the conveyance reserved the rights of the United States to the geothermal resources therein. 30 U.S.C. §1002. The act also establishes a number of exceptions to this general leasing authority for certain protected lands. 30 U.S.C. §1014.

[133] *Renewable Energy and the BLM: Geothermal*, fact sheet, available at http://www.blm.gov/pgdata/etc/medialib/blm/wo/MINERALS__REALTY__AND_RESOURCE_PROTECTION_/energy.Par. 74240.File.dat/ 09factsheet_Geothermal.pdf.

[134] *Id.*

[135] P.L. 91-581, codified at 30 U.S.C. §§1001 *et seq.*

[136] 43 C.F.R. §3203.10(a). The qualifications for nomination and bidding are set forth at 32 C.F.R. §3202.10.

[137] *Id.* at §3203.10(b).

[138] *Id.* at §3203.11.

[139] *Id.* at §3203.15(b).

[140] *Id.* at §3203.13.

[141] *Id.* at §3203.14 (a)(2).

[142] 30 U.S.C. §1004(a)(3).

[143] 43 C.F.R. §3203.15(c).

[144] *Id.* at §3203.15(b).

[145] *Id.* at §3203.17(b).

[146] *Id.* at §3203.17(c).

[147] *Id.* at §3204.5(a).

[148] *Id.* at §3204.11(a).

[149] *Id.* at §3204.12.

[150] *Id.* at §3205.6(a).

[151] *Id.* at §3207.5.

[152] *Id.* at §3207.11(a).

[153] *Id.* at §3207.12.

[154] *Id.* at §3211.11(b)(1)-(2). For both competitive and non-competitive leases issued before August 8, 2005, the royalty rate is the rate prescribed by regulation as of the date of the issuance of the lease.

[155] *Id.* at §3211.11(b)(3).

[156] *Id.* at §3211.17(a). For earlier leases, the royalty rate varies on a case-by-case basis.

[157] *Id.* at §3211.17(iii).

[158] *Id.* at §3211.18. The applicable fee schedule can be found at 30 C.F.R. 206.356.

[159] 30 U.S.C. §§1010-1011.

[160] *Id.* at §1012.

[161] *Id.* at §1010.

[162] 43 C.F.R. §3251.10. Content requirements for the Notice of Intent are set forth at 43 C.F.R. §3251.11.

[163] *Id.* at §3252.11.

[164] *Id.* at §3253.10.

[165] 43 C.F.R. Subpart 60.

[166] 43 C.F.R. §3262.11.

[167] 43 C.F.R. Subpart 3263.

[168] P.L. 94-579, 43 U.S.C. §§1767 *et seq.*

[169] *Renewable Energy and the BLM: Wind*, January 2009, available at http://www.blm.gov/pgdata/etc/medialib/blm/wo/MINERALS__REALTY__AND_RESOURCE_PROTECTION_/energy.Par.58306.File.dat/ 09factsheetmap_Wind.pdf.

[170] *Id.*

[171] *Id.*

[172] *Renewable Energy and the BLM: Solar*, July 2009, available at http://www.blm.gov/pgdata/etc/medialib/blm/wo/MINERALS__REALTY__AND_RESOURCE_PROTECTION_/energy. Par.28512.File.dat/09factsheet_Solar.pdf.

[173] *Id.* Areas of Critical Environmental Concern are defined as "those areas within the public lands where special management attention is required ... to protect and prevent irreparable damage to important historical, cultural or scenic values, fish and wildlife resources or other natural systems or processes, or to protect life and safety from natural hazards." 43 U.S.C. §1702(a).

[174] 43 U.S.C. §1761(a)(5).

[175] 43 C.F.R. §2802.10(a).

[176] 43 U.S.C. §1761(b).

[177] *Id.* A detailed review of the information that must be included in an application for a FLPMA right-of-way grant can be found at 43 C.F.R. Subpart 2804.

[178] 34 C.F.R. at §2804.14.

[179] *Id.* at §2805.15(e).

[180] *Id.* at §2805.10(a)(1).

[181] *Id.* at §2805.10(a)(2).

[182] *Id.* at §2805.10(a)(3).

[183] *Id.* at §2806.12.

[184] *Id.* at §2806.15.

[185] *Id.* at §2806.50.

[186] 76 Fed Reg. 81,906 (December 29, 2011).

[187] S. 1775.

[188] 36 C.F.R. §251.50(a).

[189] The Forest Service Chief is considered the "authorized officer," and the Regional Foresters, Forest Supervisors, District Ranger, or other officer can all serve as "authorized officers" though delegation. 36 C.F.R. §251.52.

[190] 36 C.F.R. §251.52.

[191] *Id.* at §251.53(l)(4).

[192] *Id.* at §251.54(a).

[193] *Id.* at §251.54(e)(1)(i).

[194] *Id.* at §251.54(e)(1)(ii).

[195] *Id.* at §251.54(e)(1)(iv).

[196] *Id.* at §§251.54(e)(1)(iii), (v)-(ix), and (e)(5).

[197] *Id.* at §251.54(g).

[198] *Id.* at §251.54(g)(2)(ii).

[199] *Id.* at §251.54(g)(5).

[200] *Id.* at §251.55(b).

[201] *Id.* at §2151.56.

[202] *Id.* at §251.57(a).

[203] *Id.* at §251.58.

[204] *Id.* at §251.59(a).

[205] *Id*. at §251.56(b).
[206] *Id*. at §251.64(a).
[207] *Id*. at §251.64(b).

In: Renewable Energy Development ... ISBN: 978-1-62618-764-1
Editor: Carl Polat

Chapter 4

RENEWABLE ENERGY AND THE BLM: SOLAR*

U.S. Department of the Interior, Bureau of Land Management

Solar radiation levels in the Southwest are some of the best in the world, and the BLM manages more than 19 million acres of public lands with excellent solar energy potential in 6 states: California, Nevada, Arizona, New Mexico, Colorado and Utah. State renewable energy portfolios, investment tax credits for solar energy projects, volatile oil prices, and international concern about global warming have all contributed toward public and industry interest in utility-scale solar energy development. Solar energy projects can provide significant amounts of electricity while emitting virtually no greenhouse gases, but they require large areas of relatively flat land, and some technologies use substantial amounts of water—a scare commodity in the arid climates where the solar resources are the best. At this time, there are about 80 solar energy projects, totaling some 630,000 acres, proposed on public land primarily in Arizona, California, and Nevada.

Solar energy development projects on BLM-administered public lands are authorized as rights-of-way under Title V of the Federal Land Policy and Management Act if the proposed project is consistent with BLM land use planning. The applicant is required to pay the BLM's costs in processing the

---

* This is an edited, reformatted and augmented version of the U.S. Department of the Interior, Bureau of Land Management publication, dated January 2013.

right-of-way application, and all projects require an environmental review under the National Environmental Policy Act. Any entity that receives a solar energy right-of-way authorization must comply with the terms and conditions of the authorization and pay fair market value for use of the public lands.

In 2010, the BLM approved the first utility-scale solar energy projects on public lands. To date, it has approved 11 projects that include all of the solar energy technologies considered to be commercially viable (parabolic trough, power tower, dish engine, and photovoltaic systems). These 11 projects have the potential to generate 4,512 megawatts of clean, renewable energy - enough energy to power roughly 1.3 million homes. The projects range in size from a 45-megawatt photovoltaic system on 422 acres to a 1,000-megawatt parabolic trough system on 7,025 acres. In 2011 and 2012, the BLM also approved rights-of-way for electric transmission lines on public lands that will enable the construction of six additional projects on private and Tribal land. Those projects will generate another 1,475 megawatts.

Just as with oil and gas production, not all BLM-administered lands are available for alternative energy production. Lands designated as part of the BLM's National Landscape Conservation System are not open to solar energy development. In addition, some special management areas such as Areas of Critical Environmental Concern may not be suitable for development. These areas are generally identified in existing BLM land use plans.

In October 2012, Secretary of the Interior Ken Salazar signed the Record of Decision finalizing a program for spurring development of solar energy on public lands in six western states. The Programmatic Environmental Impact Statement (PEIS) for solar energy development provides a blueprint for utility-scale solar energy permitting in Arizona, California, Colorado, Nevada, New Mexico and Utah by establishing solar energy zones with access to existing or planned transmission, incentives for development within those zones, and a process through which to consider additional zones and solar projects. The Solar PEIS establishes an initial set of 17 Solar Energy Zones (SEZs), totaling about 285,000 acres of public lands, that will serve as priority areas for commercial-scale solar development, with the potential for additional zones through ongoing and future regional planning processes. If fully built out, projects in the designated areas could produce as much as 23,700 megawatts of solar energy, enough to power approximately 7 million American homes. The program also keeps the door open, on a case-by-case basis, for the possibility of carefully sited solar projects outside SEZs on about 19 million acres in "variance" areas.

Updated January 2013.

In: Renewable Energy Development … ISBN: 978-1-62618-764-1
Editor: Carl Polat

Chapter 5

RENEWABLE ENERGY AND THE BLM: WIND*

U.S. Department of the Interior, Bureau of Land Management

The BLM manages 20.6 million acres of public lands with wind potential. The BLM's Lands and Realty Management program has authorized 32 wind energy development projects with a current total installed capacity of 581 megawatts, enough to supply the power needs of over 175,000 homes. In addition, the BLM has approved 170 wind energy site testing authorizations.

A Programmatic Environmental Impact Statement (EIS) relating to the development of wind energy on the public lands was completed in June 2005. This EIS provides an analysis of the development of wind energy projects on the public lands in the West. In conjunction with the publication of this EIS, the BLM amended 52 land use plans to allow for the use of applicable lands for wind energy development. BLM offices are able to use this EIS as an aid in analyzing impacts for specific applications for the use of public lands for wind energy use. The BLM issued a wind energy policy in December 2008 to provide guidance on best management practices (BMPs); measures to mitigate potential impacts on birds, wildlife habitat, and other resource values; and guidance on administering wind energy authorizations.

---

* This is an edited, reformatted and augmented version of the U.S. Department of the Interior, Bureau of Land Management publication, dated January 2013.

Wind power uses the naturally occurring energy of the wind for practical purposes like generating electricity, charging batteries, or pumping water. Wind turbines capture the kinetic energy in the wind, converting it into electrical energy. Utility-scale turbines are mounted on tall towers, usually 300 feet or more above the earth's surface, where the wind is faster and less turbulent. In utility-scale power applications, multiple turbines are connected to the utility grid, providing electricity when the wind blows.

For over a decade, wind energy has been the fastest growing energy technology worldwide, achieving an annual growth rate of over 30 percent. In the United States, the current total installed capacity is over 48,600 megawatts of wind projects.

Laws enacted in most of the Western states require energy companies to provide a portion of their energy from renewable energy sources (Renewable Portfolio Standard). As a result, the BLM anticipates a continued interest in the use of public lands for renewable energy development.

The BLM continues to conduct studies necessary to evaluate and process applications for rights-of-way for the siting of wind energy projects and applications for rights-of-way for electric transmission lines from these projects. The BLM has approved wind energy projects in Arizona, California, Idaho, Oregon, and Utah and, in October 2010, approved the first wind energy project on public lands in Nevada, with a potential capacity of 150 megawatts. The BLM currently has some 40 pending wind energy development applications on the public lands with a potential capacity of over 7,500 megawatts.

The BLM's wind energy program is part of ongoing efforts to improve the management of energy resources found on federal lands in a balanced way to ensure the Nation's economic and energy security and quality of life.

Updated January 2013.

In: Renewable Energy Development … ISBN: 978-1-62618-764-1
Editor: Carl Polat

Chapter 6

RENEWABLE ENERGY AND THE BLM: GEOTHERMAL*

U.S. Department of the Interior, Bureau of Land Management

The BLM has the delegated authority for leasing 245 million acres of public lands (including 104 million acres of National Forest managed by the U.S. Forest Service) with geothermal potential in 11 western States and Alaska. The BLM presently manages 818 geothermal leases, with 59 geothermal leases in producing status generating about 1,275 megawatts of installed geothermal energy on public lands. This amounts to over 40 percent of U.S. geothermal energy capacity and supplies the electrical needs of about 1.2 million homes.

In May 2007, the Department of the Interior published final regulations on geothermal energy production on public lands requiring more competitive leasing, offering simplified royalty calculations, and sharing $4 million per year in current royalties with counties where production occurs.

The BLM's geothermal leases generated over 4,600 gigawatts of electrical power during 2011 and provide alternative heat sources for direct-use commercial endeavors. These Geothermal leases generate over $12 million in federal royalties each year. Competitive lease sales since 2007 have netted over $76 million in bonus bids for geothermal lease parcels in Colorado,

* This is an edited, reformatted and augmented version of the U.S. Department of the Interior, Bureau of Land Management publication, dated January 2013.

Idaho, Oregon, Utah, Nevada, and California. A competitive auction of public lands in Nevada held in August 2008 was the largest geothermal sale ever in terms of dollars bid, bringing in a record $28.2 million for a total of 105,211 acres. The development and production of the geothermal resources on these lands will bring the Nation closer to meeting its targets for electrical energy production from renewable resources.

A Programmatic Environmental Impact Statement (EIS) relating to the authorization of geothermal leasing was completed in October 2008, and the Record of Decision was signed in December 2008. The Record of Decision amended 114 Bureau of Land Management resource management plans and allocated about 111 million acres of Bureau-managed public lands as open for leasing. An additional 79 million acres of National Forest System lands are also legally open for leasing. The completion of this document has resulted in a significant demand for new leases for geothermal resources.

Since the completion of the Programmatic EIS, the BLM has competitively leased over one million acres of federal lands, generating over $76 million in bonus bids for geothermal lease parcels. The overall lease sales include 67 parcels in Utah, totaling 241,490 acres; 11 parcels in Oregon, totaling 41,362 acres; 13 parcels in Idaho totaling 17,580 acres; 14 parcels in California, totaling 14,110 acres; 1 parcel in Colorado, totaling 799 acres; and 251 parcels in Nevada, totaling 724,085 acres.

Updated January 2013.

INDEX

F

G

H

S

T